BASS LAKE:

A GOLD RUSH ARTIFACT

BASS LAKE: A GOLD RUSH ARTIFACT

John E. Thomson, Ph.D.

Clarksville Region Historical Society
El Dorado Hills, California

Copyright © 2019 by John E. Thomson. All rights reserved. No part of this publication may be reproduced or transmitted in any form or by any means, electronic or mechanical, including photocopy, recording or any storage and retrieval system, without written permission from the author. However, no copyright is claimed for material from government publications

Although the author and publisher have made every effort to ensure the accuracy and completeness of information contained in this book, we assume no liability for errors, inaccuracies, omissions or any inconsistency herein. Any slights of people, places, or organizations are unintentional.

Contact Clarksville Region Historical Sociey, 501 Kirkwood Court, El Dorado Hills, CA 95762, website edhhistory.org.

ISBN 978-0-9669392-3-1

DEDICATION

This book is dedicated to

Betty January

whose inspiration and vision helped create the Clarksville Region Historical Society.

ACKNOWLEDGEMENTS

I would like to thank the Board of Directors and the members of the Clarksville Region Historical Society for their encouragement and support. Also those other individuals, too numerous to list, who furnished me with information, advice, and encouragement.

The Public Information Office of the El Dorado Irrigation District very generously supplied me with access to publications and documents related to the history of their organization and their ownership of Bass Lake. The staff of the El Dorado Historical Museum were responsive to my requests, and I appreciate their assistance.

I would like to extend a special thanks to Mike Roberts, who was always generous with his suggestions about style and phrasing.

I would also like to acknowledge the invaluable assistance of author Jean E. Starns, whose book *Gold From Gold Rush Waters* furnished a wealth of information about the various ditch systems of El Dorado County.

My greatest thanks go to my wife Fran, for her helpful critical reviews and her loving encouragement.

PREFACE

The inspiration to write a book about Bass Lake sprang from the news in 2014 that the El Dorado Irrigation District was offering the Bass Lake property for sale.

Up to that time I, as well as most nearby homeowners, assumed that the lake was simply a construct of the irrigation district for the purpose of supplying water to their customers.

A call to my friend and local archeologist Melinda Peak led me to discover that the lake was in fact more than 150 years old, and had figured largely in supplying water to mines and miners on the Western Slope of El Dorado County during the California Gold Rush.

The lake later became a vital component of water companies that furnished water to the farms and fruit orchards that sprang up in the 1870s, because obtaining water from sources such as Folsom Dam or groundwater irrigation pumps were decades in the future.

Further research revealed that the lake was built as the western terminus of the ditch system which later became generally known as the Crawford Ditch.

Popular opinion relegates Gold Rush activity to locations such as Placerville, Sutter's Mill, and mining towns along Highway 49. However, considerable gold mining took place around communities such as Diamond Springs, El Dorado, Shingle Springs and Clarksville.

The story which you are about to read is how water was brought to the gold miners around Shingle Springs, El Dorado and Clarksville, where water—then and now—was and is an essential ingredient of any successful mining operation.

CONTENTS

1. Introduction
2. Mining Methods ~ *Panning ~ Miners' Cradle ~ Long Tom ~ Sluice*
3. Water and Mining ~ *Ancient Peru ~ Roman Europe ~ Brazilian Gold Rush*
4. Mining Ditches in the Gold Rush ~ *Canals and Ditches ~ Flumes ~ Dams and Reservoirs ~ Evolution of the Ditch and Canal Companies*
5. El Dorado County Geography and Geology ~ *Dry Diggings ~ Ancient Stream Beds ~ Tertiary Period ~ Bringing Water to Dry Diggings*
6. Early Cosumnes River Ditches (1850-1854) ~ *Bradley, Berdan & Co. ~ Jones, Furman & Co. ~ Hard Times for Ditch Companies*
7. Ditch Company Consolidation (1855-1873) ~ *Sale of Jones, Furman & Company to Scott ~ Jones, Furman Renamed Eureka Canal ~ Diamond Springs Conflagration ~ Sale of Bradley, Berdan to Harris ~ Bradley, Berdan Assets Added to Eureka Canal ~ New Eureka Canal Company Incorporated ~ Scott Irks Eureka Canal Board ~ Miners' Water Needs*
8. Construction of American Reservoir (Bass Lake) ~ *Ditch Extended to American Reservoir ~ Evidence of Reservoir on GLO Plat*
9. Horace Greeley on Mining Canals ~ *Greeley's Advice: "Go West" ~ Greeley Goes West ~ Greeley on Ditches and Canals*

10 Eureka Canal Company Under D.O. Mills ~ *Darius Ogden Mills, Banker ~ Bank of California ~ Millbrae*

11 Park Canal & Mining Company ~ *Eureka Ditch Purchase ~ J.J. Crawford ~ Park Canal Operations ~ J.J. Crawford Family ~ J.J. Crawford Public Life ~ Park Canal Improvements ~ American Reservoir and Clarksville ~ Sale of Park Canal*

12 Diamond Ridge Water Company ~ *Formation of Diamond Ridge Water Co. ~ Transfer of Eureka Canal to Diamond Ridge*

13 El Dorado County Water Users Association ~ *Western States Expands Water Use ~ Railroad Commision Complaint Filed ~ Western States Concedes Canal Assets*

14 El Dorado Irrigation District (1925-1955) ~ *El Dorado Water Company Reorganization ~ El Dorado Irrigation District ~ New Water Supply Sought ~ Difficulties Encountered ~ Aid Sought ~ Ditch System in Need of Repairs ~ Sly Park Project Favored ~ District Asks Study of Reclamation Bureau ~ Beyond Ability to Repay ~ Proposal Goes to Congress ~ Bill Passed Over Stiff Opposition ~ A Deeper Meaning ~ Refinancing Saves Large Sum ~ Conflict Over Webber Dam Spillway*

15 Bass Lake Road ~ *Early Road Activity ~ Early Road Houses ~ Morrison House ~ Atlantic House ~ Green Springs or Dormody House ~ Delay of the Road ~ Later Road Developments*

16 James Nicol and the Bass Lake Resort ~ *James M. Nicol ~ Memories of the Bass Lake Resort ~ James Nicol in the News ~ The Walkers' Picnic at Bass Lake ~ Sale of Diamond Ridge Water Company in 1938 ~ Evidence of U.S. Geological Survey Maps ~ Sale to Monroe and Valdyne Jannke ~ Evidence of California Department of Water Resources ~ Sale To El Dorado Irrigation District*

17 El Dorado Irrigation District (1955-2015) ~ *Bass Lake and Recycled Water ~ Bass Lake Declared Redundant*

18 Rescue Union School District ~ *Bass Lake Purchased for School Site ~ Alternate School Site Purchased ~ Bass Lake Offered for Sale*

19 El Dorado Hills Community Services District ~ *Bass Lake Purchased for Park Site*

Appendix: Timeline of Significant Events

Bibliography of Sources Consulted

~ 1 ~
INTRODUCTION

Most people who live around Cameron Park and El Dorado Hills know that Bass Lake is on Bass Lake Road, a mile or so north of Highway 50. But few, if any of them know from whence Bass Lake came, or why it is there.

Old-time residents recall that at one time you could picnic and go fishing at Bass Lake. Stories abound, but to many, the lake's origins are lost.

No doubt many would be surprised to learn that Bass Lake, or American Reservoir, as it was known then, was a part of the great mining ditch systems of the California Gold Rush. American Reservoir became the western terminus of what was first known as the Eureka Ditch, later as the Crawford Ditch, and then as the Park Canal and Mining Company ditch system.

This is the story of the development of the mining ditches of southern El Dorado County, the story of the ditch companies that built those ditch systems, and how and when Bass Lake, originally known as the American Reservoir, came to be a living artifact of the Gold Rush of 1849.

Every effort has been made to ensure that the information presented in this book is accurate. However, citations have been kept to a minimum for ease of reading. Readers interested in pursuing a particular subject in greater detail are referred to the Bibliography at the back of the book.

The history and significance of water in connection with the mining of gold in El Dorado County is much overlooked today. Indeed, the icon of the Gold Rush is the miner crouched

The gold panner, icon of the Gold Rush

over a flowing stream, panning the gravel to obtain gold nuggets.

The tributaries of Carson and Deer creeks near Clarksville were worked by miners every winter starting in the fall of 1850, with rocker and tom, and paid good wages when a wet season afforded a sufficiency of water in the rivers and streams. But after 1855, the easy pickings along running water were gone, and after that, the vast majority of gold found in El Dorado County had to be wrested not from the streams

3 Bass Lake: A Gold Rush Artifact

or rivers, but from the dry areas of the county. The remaining pay dirt was concealed in the flats and heads of ravines under tons of rock.

So the miners began looking for gold in the gold-bearing sediments in ancient stream beds, the gold having been carried over many years along the ancient streams. But no water was available at these aptly-named "dry diggings." To answer this need, the water was brought to the dry diggings in man-made ditches that originated at the American and Cosumnes rivers.

These main ditches, built and maintained by the ditch companies, were tapped by the miners using smaller lateral ditches that brought the water to their claims.

The ditch companies charged the miners for the water by metering the water using a measure which was referred to as the "miner's inch." A miner's inch was generally accepted as the amount of water that would flow through a hole an inch square during 24 hours, or about 17,000 gallons of water.

Without ditch water having been brought to those otherwise dry diggings, there would have been no long-lived Gold Rush, since as we have seen, the easily obtainable gold along the natural rivers and streams was pretty well exhausted by the time the greater influx of gold-seekers arrived in California.

Western El Dorado County, from Weber Creek to the Cosumnes River abounded in good diggings, albeit dry, both on the surface and in ravines.

It was not until 1858 that the Eureka Canal Company extended its ditch system westward to the American Reservoir to serve the miners of Clarksville, Jay Hawk, Carson Creek and Western Diggings, Plunkett's Diggings, and Marble Valley.

This book is about the ditches from the North Fork of the Cosumnes River, which brought water to the southern region of El Dorado County. These ditch systems brought water from Sly Park to Diamond Springs, to El Dorado (then Mud Springs), where they ended at Bass Lake (then the American Reservoir). The spillover of Bass Lake was fed into Carson Creek, which passed through the town of Clarksville.

First used to supply the gold miners with water to wash their gold, these ditches later provided irrigation to the farmers and ranchers of this part of the county. The ditches have all but disappeared, and the only trace of their previous existance is their western terminous, our own Bass Lake.

The author would like to reassure the reader that he has attempted to the best of his ability to present the history of Bass Lake through the years in a scholarly and accurate manner.

However, in the interest of readability, I have omitted much of the specific citations and other impedimentia that usually appear in a work of history, opting instead to use a personal, rather than an academic, approach.

Should the reader wish to pursue further histories of the nature presented herein, I commend to them the reference books and publications set forth in the Bibliography, the literature upon which this book is based.

~ 2 ~

MINING METHODS

Let's review the methods that the miners used to extract gold from the gold-bearing sediments. Here we are talking about placer mining. Unlike hardrock mining, which extracts veins of precious minerals from solid rock, placer mining is the practice of separating heavily eroded minerals like gold from sand or gravel. The word placer is thought to have come from Catalan and Spanish, meaning a shoal or sand bar, and entered the American vocabulary during the California Gold Rush.

Water action is the basic mechanism by which the gold is separated from the surrounding earth and gravel, and water is essential to placer mining. Primitive waterless extraction methods, such as tossing the pay dirt in the air to separate out the heavier gold particles, have proven to be very unproductive.

In our review, we will proceed from the least efficient to the more efficient methods, examining the miner's pan, the miner's cradle, the long tom, the riffle-box or sluice, the tail sluice, and the ground sluice. (Hanks 1882, 28)

Panning

Gold panning is a manual technique of separating gold from other materials. Panning is the simplest method of placer mining. The sedimentary sand and gravel material (which

An illustration of gold panning. Men working a miner's cradle are shown in the background.

we will henceforth call "pay dirt") to be panned is usually scooped, or shoveled from existing stream beds, often at the inside turn in the stream, or from the bedrock shelf of the stream, where the density of gold allows it to concentrate, thus creating gold deposits. Wide, shallow pans filled with pay dirt are submerged in the water. The pay dirt is worked by hand to break up all compacted soil. Using a circular motion, and by dipping the pan into the water, the lighter material is worked to the top and over the edge of the pan until only the fine gold and fine sand remains.

While gold panning is the easiest and quickest technique for searching for gold, it is the least efficient, except where labor costs are very low or gold traces are substantial. In a ten-hour day, a proficient panner can work roughly 100 pans, about half a cubic yard of pay dirt.

The Miner's Cradle

The more efficient miner's cradle, which appears to have been first used in the California Gold Rush, brought the weary argonauts increased productivity over simple panning. Cradles can process two to two-and-one-half cubic yards of pay dirt in a ten-hour day. However, its use is somewhat limited by the need for frequent cleanups and by poor fine-gold recovery.

The miner's cradle is a box that stands on rockers, which allows it to be rocked like a baby's cradle. The sides are sloped off at the lower extremity like those of a coal-scuttle. The upper end of the cradle is the hopper, or riddle box, and at the bottom of the box is a piece of sheet iron perforated with holes about half an inch or so in diameter. Under the riddle box a wooden chute, sometimes covered with canvas across which

Miner's cradle or rocker

Illustration of a miner using a cradle

are nailed several transverse riffle bars, slants down away from the box towards the front of the cradle.

To work the cradle, pay dirt is shoveled into the hopper. Then the miner kneels by the cradle and ladles water over the pay dirt in the hopper with one hand, while he rocks the cradle with the other. The action of the water and the rocking motion disintegrate any compacted pay dirt in the box. The resulting mud and small gravel fall through the holes in the riddle box onto the chute and run across the riffle bars and escape at the bottom of the chute, leaving the gold, black sand, and heavier particles of gravel caught behind the riffle bars. Large rocks and gravel and other debris must be periodically removed from the riddle box.

This method requires more water than panning as the weight of water required for working a cradle is at least three

9 Bass Lake: A Gold Rush Artifact

SPANISH FLAT, 1852

Miners operating a long tom at Spanish Flat, a mining camp a few miles north of Placerville.

times that of the pay dirt washed. (West 1971)

The Long Tom

The long tom is an improvement over the rocker and employs the same methodology. It consists of a slanted wooden trough about six to twelve feet long. At its upper, or head end, it is usually about twenty inches wide and gradually widens to about thirty inches at the lower end. Its bottom is usually covered by a plate of iron to prevent wear. The sides are about eight to ten inches high and cut at a slant from the bottom upwards, and the wide end is closed by an inclined riddle of punched sheet iron, similar to that forming the bottom of the hopper of a miner's cradle.

A good supply of running water is required to operate a

Illustration of slucing operations.

long tom.

The water is poured into the top of the tom, and the pay dirt is thrown in near its head by one man while another man keeps it constantly stirred. The number of men working at a time varies from two to four, depending on the amount of pay dirt to be washed and the quantity of water available. With a long tom, two men will ordinarily handle about five or six cubic yards of loose pay dirt in a ten-hour day.

Sluice

Where running water and a grade are available, a simple sluice is generally as effective as the long tom and requires less labor. The sluice has advantages over any other system both for collecting free gold and the removal of barren dirt in an economical manner. Consequently, the attention given to its construction and the work it performs will prove remunerative.

The sluice is generally a long wooden trough similar to

a long tom, having a considerable inclination or slant, into which the pay dirt is shoveled and through which a rapid stream of water continually flows. The bottom of this trough is provided with a series of riffles, usually containing mercury, by which the gold is retained, while the clay, sand, and gravel are carried off by the force of the water current.

The quantity of water available will influence the scale of operations and the size of sluice used. A minimum flow of 170 to 225 gallons per minute is required for a 12-inch-wide sluice-box with a steep incline. A sluice with a rapid current of water needs to be made longer than one that is set more nearly on a level. For this reason, where a long sluice cannot be employed, the inclination must be diminished. In general, a large body of water and a rapid current are essential.

Tail sluices are arrangements for collecting gold still retained by the clay, sand, and gravel that has passed through the ordinary sluice. They are usually placed in a ravine through which the tailings of clay, sand, and gravel from one or more ordinary sluices flow. They are only gleaned for gold every several weeks, in the meantime receiving no further attention than is necessary to prevent their becoming silted up.

In localities where there is a large supply of water, plenty of pay dirt of low yield, and the necessary slope, a sluice is sometimes improvised without the use of wood. Such arrangements are called ground sluices. In order to prepare one, a small gutter or shallow ditch is dug with a sufficient slope, through which the pay dirt is to be washed. A stream of water is directed into the ditch, the erosive action of which rapidly deepens and enlarges the channel. As soon as the sides and bottom of this ditch have ceased to become rapidly eroded by the action of the current alone, the miners begin to assist the operation by shoveling in chunks of earth from the bank, which, falling into the stream, are acted on precisely as in the case of the ordinary board sluice. No mercury or riffles are employed in the ground sluice. When a considerable amount of dirt has been passed through a sluice of this description, the water is diverted, and the remaining gold-bearing material is collected and processed for the residual gold in a tom, cradle, or short box sluice. This type of operation is sometimes referred to as "gouging" or "booming."

In his 1885 book on mining camps, Charles Shinn states that "with rich virgin soil, a ten or twenty feet frontage [along a stream] was a large claim; but the same gulch when being worked over for perhaps the fifth time, for what predecessors had been unable to obtain, was divided up into claims of one or two hundred feet frontage. The ground, worked at first by placer-miners with pick and pan, was again sifted and searched by rocker and long-tom process; then, perhaps, by ground-sluicing; and lastly, by various forms of the hydraulic process, the entire gulch, where dozens of small claims had once existed, passing under one ownership."

One can see that each of these mining methods requires the extensive use of water to separate the gold from the surrounding material; therefore the presence of a readily available water source was important to any placer mining activity carried on in California.

~ 3 ~
WATER AND MINING

Water is the great resource that aids the gold miner. The operations of the miner can be cheaply and rapidly conducted with a large supply; but without water, or with only a limited amount, a claim that would otherwise have been highly productive may either become valueless or only capable of yielding very minimal returns.

At the onset of the gold rush, a few lucky miners spotted the occasional nugget lying on the ground or in a stream, but from the start, any serious mining involved extracting ore-bearing soils or gravels and using water to separate the gold from the chaff.

Early claims along the richest streams quickly played out. Industrious argonauts followed lode-bearing veins farther and farther away from the played-out streams and soon discovered a point of diminishing returns. The work was backbreaking enough. The value of any claim was directly related to the availability of water. To this day, moving water remains the most effective tool to separate gold from surrounding soils, but techniques vary now just as they did then.

The methods employed by early California miners were the culmination of practices developed thousands of years ago in South America and also in Europe.

Ancient Peru

Peruvians mined placer deposits in Andean rivers as early as 1,200 B.C. using panning techniques that can still be seen along the American River in Coloma.

Peruvian gold was very pure and helped ancient civilizations thrive over the subsequent 2,500 years. Artisans developed highly sophisticated metallurgy and goldsmithing practices along the way, many of which were adopted by the conquering Incas in the 14th century. The results became booty to conquering Spaniards a century later. Much of that knowledge made its way north in the mid-1800s.

Roman Europe

Roman historian Pliny the Elder (A.D. 23 - A.D. 79) authored the 37-volume pre-encyclopedia *Naturalis Historia*, which purported to cover the entirety of ancient knowledge, including Roman gold mining practices. He emphasized the importance of water and revealed another use of the famous Roman aqueducts, which, along with canals and ditch-

Pliny the Elder, as portrayed by a 19th-century artist.
No contemporary depiction of Pliny is known to survive.

es, brought mountain water to the mines for the purpose of washing gold from the sedimentary debris.

Translated from Latin, Pliny described the effort: "Valleys and crevasses had to be united by the aid of aqueducts, and in another place impassable rocks had to be hewn away and forced to make room for hollowed troughs of wood; they take levels, and trace with lines the course the water is to take; and thus, where there is no room even for man to plant a footstep, are rivers traced out by the hand of man." (Bostock and Riley 1857, 99)

Brazilian Gold Rush of 1693

In 1849 English Geologist David Thomas Ansted described the 1693 Brazilian gold rush in the widely circulated "Gold Seekers Manual." Brazilian miners dug pay dirt from pits and carried it to the nearest water source, where they separated the gold in wooden vessels, "broad at the top and narrow at the bottom," by introducing water and shaking it from side to side, "till the earth was washed away and the metallic particles had subsided," according to Ansted.

The technique, perhaps a precursor to the Miner's Cradle, purportedly yielded nuggets up to twelve ounces in weight.

The manual followed Brazilian mining into the following century, when Brazilian miners realized the efficiencies of using ditches and wooden troughs to move the water to the pay dirt, rather than vice versa.

Early California miners quickly assimilated the accumulated gold mining knowledge of the day and then proceeded to adopt and improve on the techniques and equipment to suit their circumstances.

~ 4 ~

MINING DITCHES IN THE GOLD RUSH

Miners needed water to recover the gold from their pay dirt. As the streams played out they were forced to venture ever farther from their water source. Carrying pay dirt to the nearest water supply added a lot of extra work to an already arduous profession.

Enterprising miners found ways to tap into streams or even divert them completely. Enterprising former miners formed ditch companies and found selling water to miners, or stock to shareholders, more lucrative than mining.

The terms "canal" and "ditch" are largely interchangeable, with "canal" typically denoting a more significant ditch, often the "trunk" in a ditch water delivery network.

Canals and Ditches

At least four-fifths of the gold captured was obtained using ditch water. That fact alone gives canals and ditches an important place in El Dorado County history.

The Gold Country is a dry place with relatively few natural springs. The bed rock is typically slate with perpendicular cleavage, which lets surface water soak down to the water table rather than bubble back up as spring water.

Permanent waterways, such as the main tributaries of the

American River in the north and the Cosumnes in the south, are few and far between. They run generally east to west in deep, steep channels, which greatly limit access.

The vast snowfields at the headwaters of both major river systems, combined with vast mountain lakes that make ideal reservoirs, make El Dorado County's water supply the envy of neighboring counties.

Indeed, under current federal and state water supply management policies, much of the water that originates in El Dorado County passes through to become drinking water in downstream locales, or salt water dilutent for delta smelt.

The descending topography of the western slope allowed ditches to provide plentiful water supplies to all but a small portion of El Dorado County.

An example of a typical mining ditch. Note the sloped sides of the ditch and the pathway for the ditch tender.

Ditches generally carry water downhill from reservoirs. Ditch slope was critical. Fast water put undue strain on the often fragile ditch networks, either washing away the clay that lined ditches that were below grade, or damaging flume system, with undue weight, friction, and lateral forces.

The size of a ditch was determined by its capacity requirements. More capacity meant more water to sell downstream

Here a flume conducts water along a steep rocky mountainside where digging a ditch was impractical. Note the wood construction and the plank walkway laid on top of the flume for the ditch tender.

An example of a hanging flume. Note that at the time the photograph was taken the flume still carried water.

at rates that were roughly determined by the water companies using the "miner's inch" measure of water usage. The ideal downhill slope was 10-12 feet per mile, but some lower capacity ditches operated successfully at slopes up to 20 feet per mile.

The ditches were dug in the earth, and the excavated soil was piled along one or both sides of the cutting to help con-

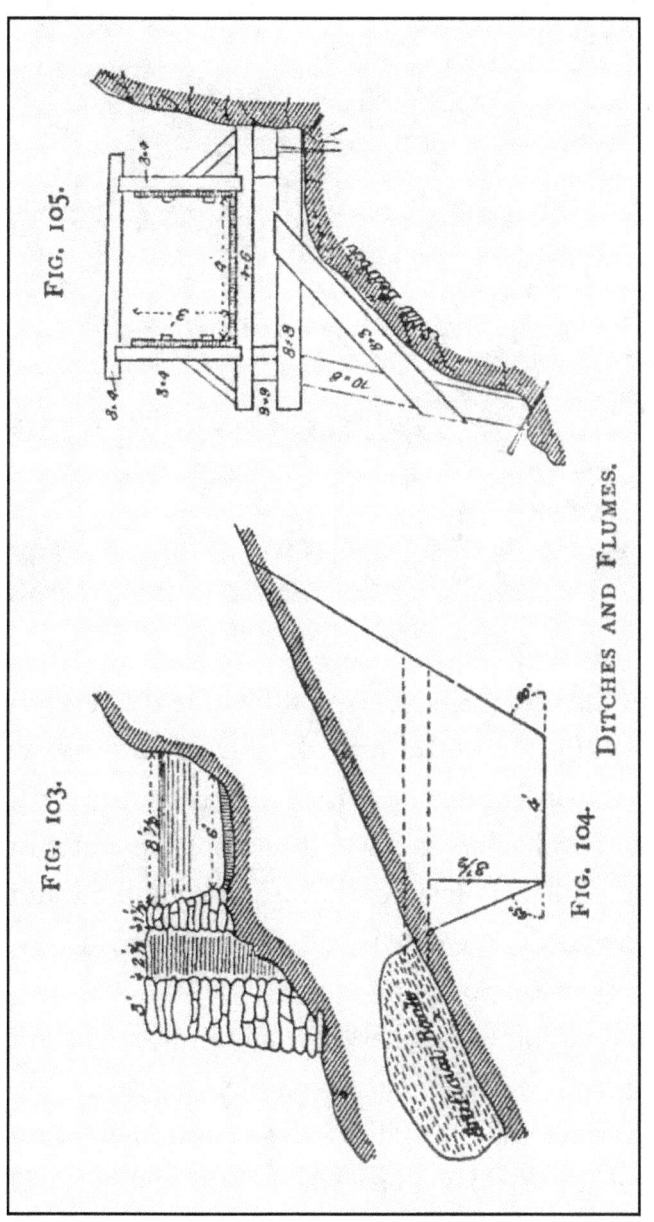

Cross-sectional examples of typical ditches and flumes: Fig. 104 shows the ordinary mode of constructing an earthen ditch along a hillside. Fig. 105 shows the method of posting along cliffs, where the foundation is occasionally narrower than the flume. Often the nature of the ground will not permit the outside bank, in which case masonry may best be substituted, as shown in Fig. 103 (from Lock, 1882).

Here a flume carries water across a stream.

tain the water in the ditch. Since most of the time ditches followed the natural contour of the land, the excavated dirt was usually piled on the downhill side of the ditch. The loose earth was packed down and acted as a wall that kept the water in the ditch. If the earth was friable, that is, it crumbled and washed away too easily, the ditch wall was lined with clay. This practice was called "puddling."

Importantly, ditch networks had to maintain a consistent slope over their course. Whenever possible, engineers followed the existing terrain along the sides of canyons. Hundreds of miles of these old waterways are still visible in El Dorado County.

The steep canyon walls wouldn't always relinquish the necessary toehold for an in-ground ditch. Those unforgiving locations required a flume.

Flumes

Flumes carried the water across valleys or in places where it was impractical to dig a ditch. For example, a flume can carry water over rocky ground, where the water must be conveyed along the face of steep mountainside or vertical cliffs, or when the water must cross over a canyon or a stream. Flumes were usually made of two-inch wood planks, or sometimes iron pipe was used.

There were also certain conditions of the formation of the ground, independent of the topography, where a ditch could not be employed so economically as a wooden flume, such as when the ground is composed of either very hard or very

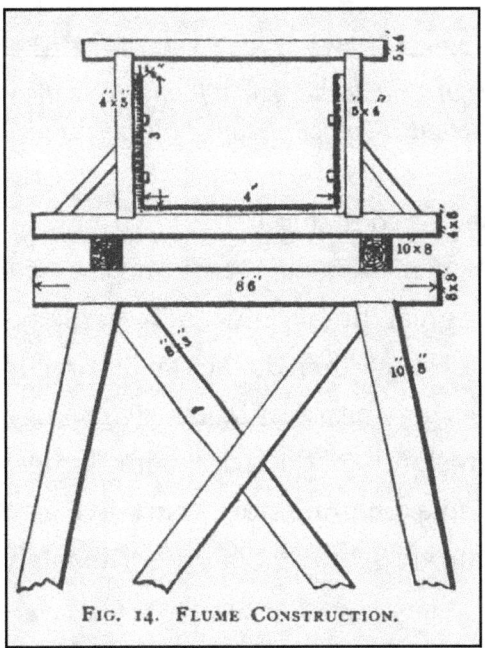

FIG. 14. FLUME CONSTRUCTION.

Cross-section view of a typical flume.

Example of a modern wooden flume. Note the walkway for the ditch tender on top of the horizontal ties.

porous and shattered material.

Likewise where water is scarce and the evaporation and absorption are great, flumes must necessarily be preferred. In such cases as these, earthen ditches are impractical, and either flumes or pipes may be advantageously used.

Because erosion is not a factor with flumes, they are set, where practicable, on grades of twenty-five to thirty-five feet per mile and are consequently of proportionately smaller area than ditches.

In general, the use of flumes was avoided wherever possible because of their high maintenance and relatively short life. In the case of flumes that carried water over obstacles, the higher a flume was built, the more apt it was to be blown down by storms. Above-ground flume structures required engineering, materials, and assembly skills far beyond hand-dug ditches.

Dams and Reservoirs

Of course, to get water into the ditches and flumes, and from there to the miners, there had to be a means of collecting the water in the first place. Varying stream levels and the inherent seasonality of natural flows limited the usefulness of simply tapping existing rivers and streams.

The extensive ditch networks required a consistent, reliable water supply. For that they needed dams and reservoirs.

The earliest dams constructed to divert water for river and placer mining were simple structures of earth and stone. In the late 1870s, water and mining companies built more extensive earth, timber, and stone dams to store water. In remote locations of the mining country, rock-fill dams were often constructed using locally obtained rock as the main structural material, although wood was often used as cribbing or to line the dam's upstream face.

In addition to high-elevation storage reservoirs, ditch companies built temporary storage, or regulating, reservoirs near the point of use. The main storage reservoirs would catch the water high in the Sierra Nevada during winter and spring and distribute the water throughout the rest of the year. Nearer to the mining area, companies had smaller distribution reservoirs. From these reservoirs, water could be easily conveyed to mining claims even if the canal system was out of service, or the reservoir could be used to retain surplus water coming from the main ditch when the claims were shut down.

At the upper end of the system, storage reservoirs were built to capture the water from small streams, natural runoff, and melting snow. Massive quantities of water were being captured by the turn of the century.

Evolution of the Ditch and Canal Companies

The first water companies emerged as early as 1850, carrying water to all the principal placer districts. These small water companies, like river mining companies, were joint stock companies formed by miners and local merchants.

The first ditches dug by these early water companies were short and relatively easy to construct. By pooling resources, the water companies could make these efforts possible.

In the 1870s, the systems that delivered water to the main mining districts of California were more difficult and complicated to build than the small mining ditches scratched out between a creek and claim in the early days of the gold rush. The earliest ditches were constructed with the goal of bringing water to where the digging was easiest.

Larger mining canals with their bigger storage reservoirs and extensive ditch systems called for skills and techniques of construction beyond the capability of most practical miners.

Through a process of canal building and consolidation, by 1899 this water was being captured by three great ditch systems on the three great ridges, or divides, within the county: the system of the California Water Company, on the North Divide; the El Dorado Canal and Deep Gravel Company's system, on the Middle Divide; and the Diamond Ridge, or Park

Canal system, on the South Divide.

However, we will discuss more of that later, as the development of the great ditch systems of El Dorado County took many years to construct.

~ 5 ~

EL DORADO COUNTY
GEOGRAPHY & GEOLOGY

As previously mentioned, bringing water to the dry diggings was necessary to separate the gold from the pay dirt. At least four-fifths of the gold recovered in California was recovered from dry diggings and done so directly or indirectly with the assistance of mining ditch water (Browne 1868, 179). Why was the gold found in these dry diggings? Most Gold Rush images show a gold miner panning for gold along a running stream or river. That image is certainly true for the first year or so of the Gold Rush, when gold nuggets were panned out of the existing streams and rivers where they lay in the stream bed; however, by 1850 or so most of these easy pickings were gone.

Dry Diggings

Realizing that the gold that had been deposited in the river and stream beds must have come from somewhere, miners sought the source, or mother lode, of the gold that had been washed downstream. One alternative was to look for gold in riverbeds where water once ran but which now were dry. An ancient riverbed could be simply the previous bed of an existing river that has changed its course over time due to flood or landslide. These old riverbeds are usually higher than the exisitng watercourse due to the erosion of the riverbed over time. In order to wash the pay dirt, either the dirt must be taken to the water or the water must be taken to the pay dirt.

Ancient Sream Beds

Another place to look for gold was in ancient stream beds that were formed in the Tertiary period.

Tertiary is a term used to describe the geologic period from 66 million to 2.58 million years ago, a time span that lies between the Secondary period and the Quaternary.

In the early 1800s a system for naming geologic time labeled only four periods. They were named using the Latin forms of numbers for first, second, third and fourth. The word tertiary means "third." It was the third period in this system. Today, we use a different system, but the name Tertiary is still commonly used for the first part of the Cenozoic Era.

Tertiary Period

The Tertiary period began with the demise of the non-avian dinosaurs and extended to the beginning of the Quaternary glaciation at the end of the Pliocene Epoch.

The Tertiary saw the demise of the dinosaurs and the rise of the mammals. The first hominids appeared in the latter part of the Tertiary.

During a great part of the Tertiary there was a vast and different system of rivers, which generally flowed southward. These rivers existed for millions of years, during which considerable amounts of gold from the rich lodes of gold were washed down into their riverbeds.

At the end of the Tertiary period, the mountain range

Bass Lake: A Gold Rush Artifact

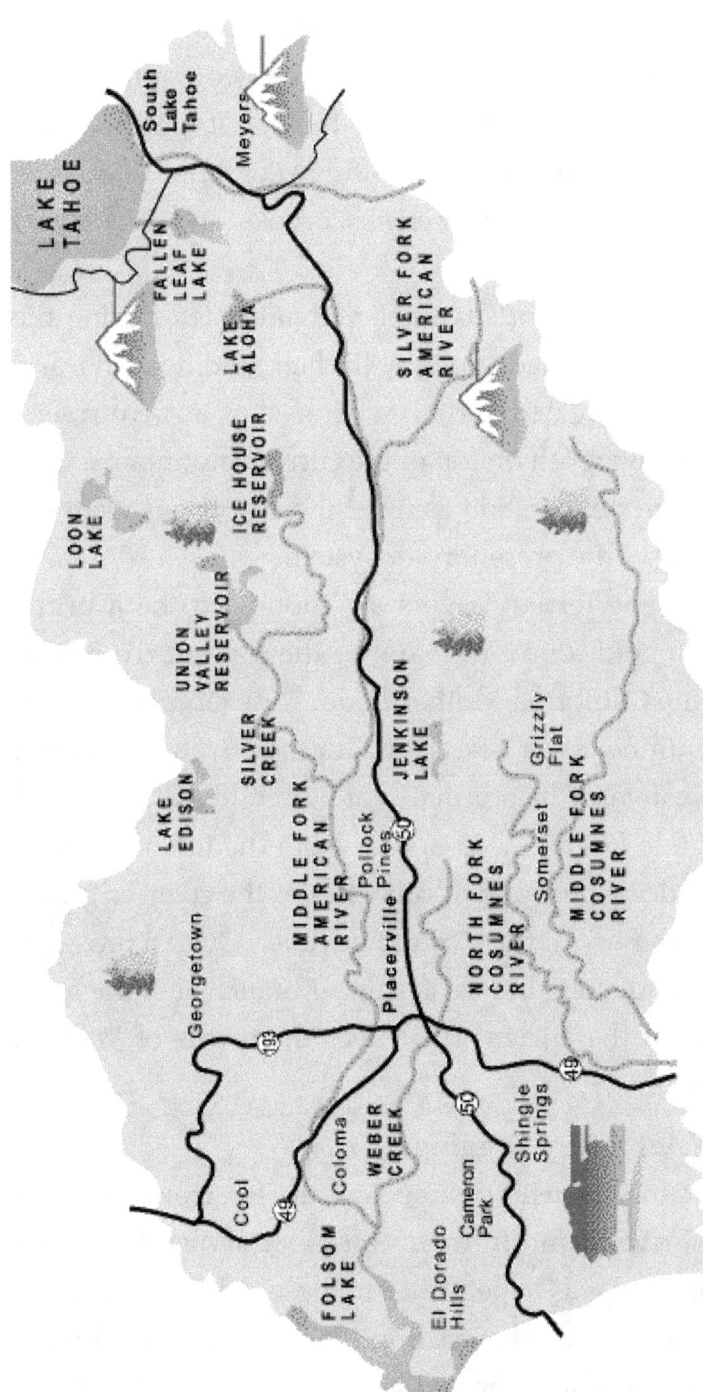

A map of El Dorado County showing the present major river systems. Note that both the American and Cosumnes river systems run generally east to west in canyons between the major east-west mountain ridges that traverse the county west of the divide. Map courtesy of El Dorado County.

system in what is now the western United States underwent drastic geological shifts that changed the geography of the Western Slope of El Dorado County to what it is today. As a result of this upheaval and the resulting geographical change, the present westward-flowing creeks, brooks, and rivers were formed.

However, the ancient streambeds that existed during the Tertiary period remained, along with their gold deposits, and that is why they created Tertiary channels or ancient rivers. Some portions were left on top of the current mountains. Others ended up in deserts, and some of these portions were left next to or part of the present river system.

These ancient Tertiary rivers are thought to be a prime source of the gold found in many of the present rivers and streams of the California Mother Lode. For example, the Tertiary Mokelumne River starts in Eldorado National Forest and runs westward. It terminates about eight miles west of California State Highway 49 not far from the town of Plymouth in Amador County. Two branches of the river originate near the North Fork of the Cosumnes River. Quite a few gold sites were found along the westernmost branch and along the main course of the ancient river near the towns of Volcano and Plymouth.

Bringing Water to Dry Diggings

Many of the ancient riverbeds were far from the new streams and rivers, and in many cases far above them. The task of the miners was to get water to these old riverbeds. The solution lay in tapping the existing rivers and streams by using ditches and canals to bring the needed water to these dry diggings.

While smaller and shorter ditches often drew their water from nearby small streams or creeks, the large ditch or canal systems that served the miners of El Dorado County may be generally classified into two general sections, depending on the source of their water: (1) those that drew their water from the American River watershed in the northern part of the county, and (2) those who obtained their water from the watershed of the North Fork Cosumnes and its tributaries.

An examination of a topographical map of El Dorado County will reveal that there are three distinct main mountain ridges running east to west: the first from the junction of the North and South forks of the American River; the second from the mouth of Weber Creek; the third from the plains between the South fork and the Cosumnes River. All originate at the crest of the water-shed.

In the northern part of the county, the Middle Fork of the American River has numerous branches, such as the Rubicon and Pilot Creeks, all having their sources among the snow of the summit range.

Even further to the south, the South Fork of the same river draws its supply from Blackrock Creek, Greenwood Creek, Rock Creek, and Silver Creek on the north, and Weber, Plum, Mill, Alde, and Alpine Creeks on the south.

Even further to the south we have the Cosumnes River, with its various forks and tributaries forming a complete network over the southeastern portion of the county. In the mountains are numerous lakes ranging in area from a few acres to many square miles, most of them so situated that, at a

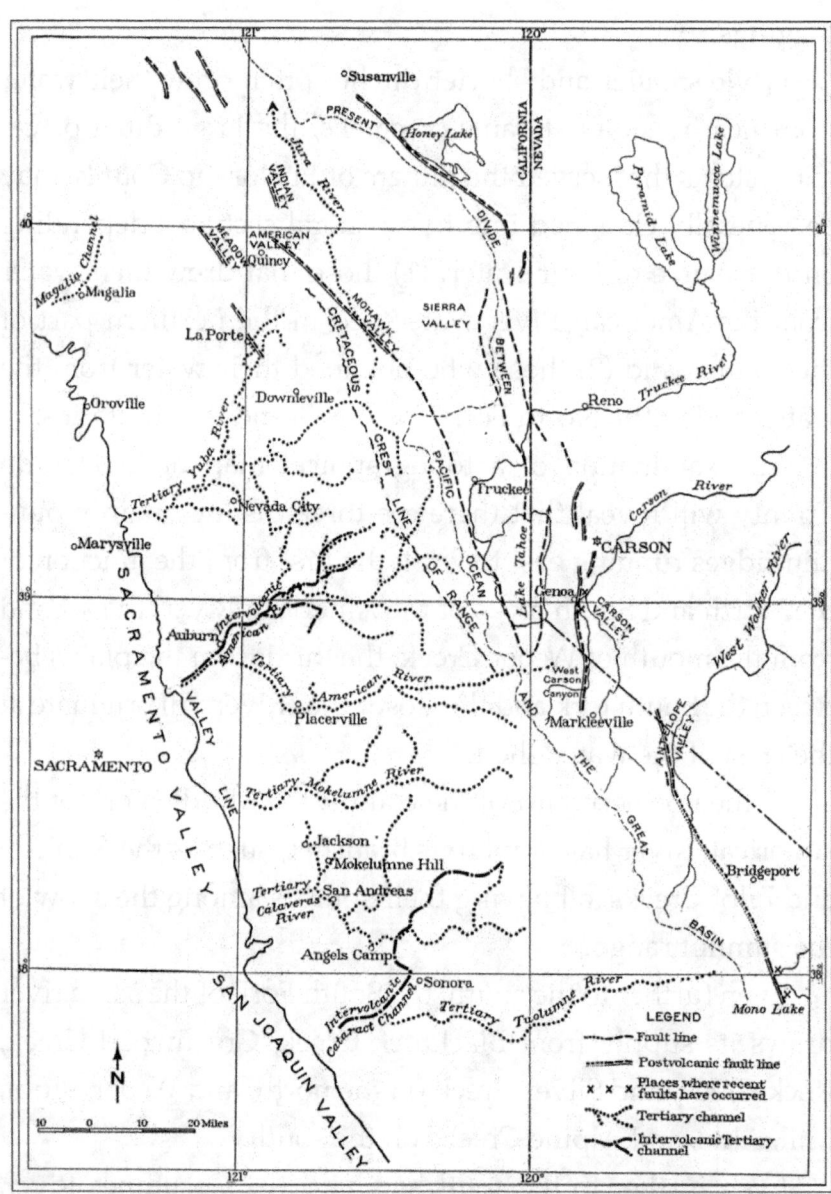

Lindgren's (1911) map of ancient Tertiary-era gold-bearing river channels in the northern Sierra Nevada, illustrating his view that the paleovalleys headed west of the present Sierra crest. Later research has challenged some of Lindgren's conclusions, but in general his early work is recognized as ground-breaking research.

small expense, they can be made useful as storage reservoirs for the great ditches below.

Tapping each of the rivers mentioned were one of the three principal canals of the county: the California Water Company, the El Dorado Water and Deep Gravel Mining Company, and the Park Canal and Mining Company. Besides these, there are numberless minor ditches, mostly constructed for minor purposes, but many of them of considerable length and importance.

This book fcouses on the ditch or canal systems that drew their water from the watershed of the North Fork of the Cosumnes River and its tributary streams in the southern part of El Dorado County. In the next chapter we will learn more about those systems.

~ 6 ~

EARLY COSUMNES RIVER DITCHES (1850-1854)

Two pioneer ditch companies figured largely in the early development of mining ditches fed by the Cosumnes River: Bradley, Berdan & Co. and Jones and Furman & Company.

Bradley, Berdan & Co.

The firm of Bradley, Berdan & Co. was among the first to take hold of ditch construction on an extensive scale. The company was incorporated on August 4th, 1851, and claimed water from Ringgold Creek and the Cosumnes River and its northern branches.

The corporation was formed by Dr. Leverett Bradley, a civil engineer from La Porte, Indiana; his son, Joseph H. Bradley; and a man named John Berdan, of whom nothing seems to be known (Starns 2004). The names of Darius Ogden Mills (whom we will meet again later) and John Parrott appear in the list of early investors.(Sioli)

Hutchings' Illustrated California Magazine was a magazine produced in San Francisco by its publisher and promoter, James Hutchings, between 1856 and 1861. The magazine played an important role in popularizing California in general, and to a large extent, Yosemite National Park.

Although Yosemite was prominent, Hutchings' magazine focused on California's many nascent tourist attractions. Each issue contained travel narratives, ranging from simple day trips out of San Francisco to arduous trans-Sierra treks. Longer articles were interspersed with shorter and lighter pieces, such as poetry and tables of interesting facts.

The *Hutchins Califonia Magazine* carried an article, "Saw Mill Railroad," in their August 1857 edition describing the saw mill that Dr. Bradley (of Bradley, Berdan & Co.) constructed on the North Fork of the Cosumnes River. The following narrative is extracted from that article.

Illustration of the saw mill railroad from Hutchins Magazine article.

The preceding illustration, representing a Saw Mill Railroad, constructed on the side of a steep mountain, on the north fork of the Cosumnes river, near Sly Park, shows what can be done to accomplish a given purpose, when it is required. In the summer of 1852 this railroad and a saw mill were erected in this wildly romantic spot, under the superintendence of Dr. Bradley of the corporation or Bradley, Berdan & Co., for the purpose of sawing the lumber required in the construction of their large canal, from this stream to the mining towns of Ringgold, Weberville, Diamond Springs, Missouri Flat, El Dorado City, (then called Mud Springs,) Logtown, and several other mining localities in the southern portion of El Dorado county, to supply those districts with water for mining.

This railroad is built upon an inclined plane, at the (often quoted) angle of forty-five degrees, for the purpose of lowering saw-logs to the mill. The car descends with its load, and being attached by a rope thro' a pulley at the top to the empty car, the weight descending causes the empty car to ascend; and by which contrivance the necessity of any other kind of machinery for that purpose is obviated.

We happened to be one of a very agreeable little party to visit this singular place, and could the reader have seen us—ladies and gentlemen, cold chickens and sandwiches, boiled ham and water melons, blankets and daguerrean instruments—all snugly stowed away in that coach, and then have heard the jokes and fun going on, if he had not been envious of our enjoyment, we know he would like to have been of the party,—that is, if he liked pleasant company.

On, on, we go, as merry as crickets; now passing through long forests of trees; now ascending or descending a gently rolling hill; then taking alternates doses of dust and soda water—jokes and cakes—until we arrived at the top of a hill overlooking a canon. Here, on looking down, we saw something resembling two long lengths of broad ribbon with bars across, lying on the side of the hill. When the question was asked, "What is that?" it was answered with "that is a railway, and we take all our logs down that rail to the mill—that dark spot down yonder; and we have all to take a ride on it to the mill."

After some hesitation and delay, one gent seats himself in the car, (fitted up with seats for the occasion,) and with sundry questions and entreaties, and sighs and oh dears, the whole party join him, and at last we are all safely seated ; while beneath the seats are the water-melons and blankets, cold fowl and daguerrean instruments, cakes and shawls, pies and over-coats. Now off we go!

Slowly we started, and with many heart flutterings and tremblings, fears and exclamations, on, on, we go, until the anticipated danger over, we all stand in safety at the bottom of the railway; and then we calmly looked our enemy in the face and took courage.

Sundry other remarks of surprise and consolation, were interrupted by our guide and host, Dr. Bradley, who informed us that the perpendicular height of the hill from where we stood to the top, was seven hundred feet, and the length of the railway on the steep side of the hill, was only one thousand feet in length.

"You saw the building at the top, where the logs lie?" Bradley continued.

"Yes."

"That is called by the workmen the 'hypo,' and the

mill down here where we stand, they call the 'depot.' Just look around."

We did look around, but what a wild, craggy place for a mill, that itself was built upon rocks; the fire-place, hearth and chimney in the kitchen were all natural formations of the rock. A flume which has been constructed, is built, or rather hung upon rocks; a prop here, a packing there, and a brace yonder; here, a tree cut off, formed a post; there, a rock formed a stay; while the water rushed and leaped on, on, down the sleep rocky bed of the river, as though it cared for nothing and no one.

Friend Bradley we give you credit for your undaunted perseverance. This work, with many others, shows what can be accomplished by patient, unswerving determination and skill. If at any time a miner should, for a moment, be disposed to think lightly or water companies, we wish him to visit the upper end of most of our canals, there to witness the expense, labor and energy expended on them. At this mill was sawed all the lumber needed in the construction of the flume; besides supplying many thousands or feet of lumber, for sluice making and other purposes, in the settlement below.

It is a magnificent sight to see the stately pine and venerable oak, growing upon and among vast piles of rocks, in some instances a large overhanging tree growing in the seam, or between two rocks; as though it were a lever placed there by nature to overturn portions of the mountain above, adding wildness, boldness, beauty and sublimity to the beautiful landscape.

After enjoying the good things provided by our worthy host, and all the pleasant and exhilarating recreations of fun and frolic, we wended our way along a plank on the top of a serpent-like flume, until it intersected the road below, (as none cared to ascend that railway again,) where our coach had been sent to meet us, and soon we were "all aboard," and on our way homeward, indulging in the reminiscences and enjoyments the trip had afforded us. Should any of our readers ever go upon a jaunt of this kind, they have our best wishes that an equal amount of pleasant und sunny gladness may keep them company on the way, and then we know that they will say, "Yes, we enjoyed it," when the journey is ended.

At some time around or after 1852 Bradley, Berdan & Co. completed the Ringgold Ditch, tapping the creek of that name; the South Fork of Weber Creek; and the Bradley Ditch, taking up the waters of Sly Park and Camp Creeks. Distributing ditches spread the water all over the Diamond Springs and Mud Springs area (Sioli 1883, 105).

The company was forced to obtain a mortgage for $18,000 from the Bank of D.O. Mills in October of 1855 and obtained another mortgage in June 1857 to pay creditors for the construction of their ditch. (Starns 2004)

Jones, Furman & Company

Historian George Peabody reports that Jones, Furman & Co. were digging their Diamond Ditch from Diamond Springs to a weir on Squaw Hollow Creek (below Oak Hill Road) by 1851. (Peabody 1988)

Peabody describes the manner of work: "Miners, who were unable to to find a wage-paying claim or whose claims were too far from water to be worked in the summer, joined the company. Work was done with picks, shovels and sweat. Rocks were moved with crowbars, and they were broken with sledge hammers, wedges and black powder (dynamite would not be invented for 15 years). Machinery and work animals were in short supply." (Peabody 1988)

Describing the Diamond Ditch, Peabody says that the Diamond Ditch followed the contours of the hills at a steady incline (about 12 feet to the mile, except where a creek bed was used), assuring a substantial but controllable flow of water for the gold miners. (Peabody 1988)

By 1852, Jones, Furman & Company had extended their ditch eastward from Squaw Hollow Creek to Clear Creek, then eastward again to their dam on Camp Creek by 1853.

An article about Jones, Furman & Company that appeared in the Sacramento Daily Union of November 15, 1853, stated that "the present company consists of M.K. Spearer, (President,) Zalmon B. Furman, Tyra B. Harris, Alfred M. Jones, A.H. Hawley and Mr. Carpenter."

Jones, Furman & Company became a corporation in early 1854. The owners of the company transferred their interests to the corporation on January 16, 1854. (Book A, Page 469) On February 20, 1854, the Sacramento Daily Union reported that "The Water Company of Jones, Furman & Co. has been incorporated, and the capital stock increased to $350,000, of which $50,000 have been sold at par value since January 16th last." On April 22 of that same year, an article in the Daily Union reported that the officers of the corporation were A.M. Jones, President; George McKenzie, C.G. Carpenter, and A.M. Jones, Trustees.

Hard Times for Ditch Companies

The years 1854 and 1855 saw the beginning of a decline in the fortunes of the ditch companies. There were several reasons for their financial deterioration: the falling price of water, the expensive cost of construction, and the decrease in the yield of the gold mines, which led to miners being unable to pay for the water they consumed. Ditch owners often found themselves saddled with unprofitable investments.

The first experiments in ditching in 1850 were magnificently successful. The canals were short and small, and the water was either sold at a very high price or was used in working out rich claims. It was not uncommon for several years for little ditches to repay the cost of construction in a couple of months. Numerous ditch companies were formed to bring water from the elevated regions in the mountains, and many invested too much for them to withdraw before any of them had learned the business.

In 1851 and 1852 the common price for water was 50 cents or $1 per miner's inch, and the ditch companies made their calculations upon charging those figures, but before the completion of the ditches the best claims in the ravines had been exhausted, and there was not enough rich ground left to pay high prices for all the water. A miner's inch is roughly the amount of water that would escape in twenty-four hours from an aperture one inch square through a two-inch plank with a steady flow of water standing six inches above the top of the escape aperture, which equates to about 2,274 cubic feet, or 17,000 gallons. (Starns 2004)

Almost without exception, the big ditches proved unprofitable. It was estimated by competent men that by 1868 not less than $20,000,000 had been invested in the mining ditches of California and that their cash value at that time was not more than $2,000,000. In many cases they broke the men who undertook them. Most of them were sold by the sheriff, some of them several times over.

There was a steady decline in the value of the ditches because there was a steady decrease in the yield of the mines, which consumed nine-tenths of the water. It was estimated in 1868 that the receipts of the ditches in California decreased ten percent a year on an average, while there is no corresponding decrease of expenditures.

By 1854 and 1855, many of the companies were seriously troubled by an inability to sell all their water, and some had commenced to buy up mining ground to wash on their own account.

Frequently miners found their claims would not pay, and owing a debt to the water company for the water they had used, simply transferred their claims to the water company in payment. To work these poor claims the company hired Chinamen, required the ditch tenders to devote their spare hours to the labor of superintendence, and by using water for which there was no market, at times managed to make a profit where the original claim owners could make none. (Browne 1868)

However, these and other deficiencies eventually engulfed first Jones, Furman and Company, and later, Bradley, Berdan & Company.

Next we shall examine how these companies were combined to form a large viable ditch company that served southern El Dorado County with water from the North Fork of the Cosumnes River.

~ 7 ~

DITCH COMPANY CONSOLIDATION

There are several accounts as to how Jones, Furman & Company was combined with Bradley, Berdan & Company.

Peabody states that by 1854 Jones, Furman & Company had completed the North Fork Canal to the weir near the Steely Fork of the North Fork of the Cosumnes River. Peabody then says that, the firm being in financial difficulty, its properties were sold at a sheriff's auction to W.P. Scott. (Peabody 1988)

Sale of Jones, Furman & Company to Scott

Historian Paolo Sioli states that in 1854, the Jones, Furman & Company property was purchased at a sheriff's sale by W. P. Scott, then of Diamond Springs, who renamed it the "Eureka Ditch." Sioli says that Scott extended the ditch to the North Fork of the Cosumnes and took up Steeley Fork. (Sioli 1883, 106)

In any case, the Placerville *Mountain Democrat* of July 8, 1854, carried an announcement of a sheriff's sale in execution of a judgment of May 9, 1854, against defendants H. Jones, A.M. Jones, Z.B. Furman, and others. El Dorado County Sheriff David E. Buel announced the sale of property he had seized as a result that consisted of all the right, title, interest, and claim of the the defendants of, in, and to, the flume, canal, reservoir, and other works belonging to the firm of Jones, Furman & Co., situated in the county, commencing at the headwaters of the Cosumnes River, and extending to the village of Diamond Springs, and other mining localities in the county. It appears that the sheriff was eventually able to sell the property to William P. Scott, who bought Jones, Furman & Company from him for $65,000.00 on January 8, 1855. (Book B, Page 291; Starns 2004)

Jones, Furman Renamed Eureka Canal

Under Scott's ownership, Jones, Furman & Company was renamed the Eureka Canal Company. Scott extended the canal to the confluence of the North Fork of the Cosumnes and Steely Fork. (Starns 2004)

Diamond Springs Conflagration

In 1856, the town of Diamond Springs was almost entirely burned down. According to contemporary reports, at about nine o'clock in the morning flames were discovered issuing out of the Howard House, a large building in the heart of town and built of the most combustible material, which soon spread above, below, and across the street, sweeping everything before them. There was a strong breeze at the time, which carried the sparks in every direction, and increased the fury of the fire. The water from the Eureka Canal was turned down into the town and assisted materially in checking the flames. About eighty buildings were destroyed, involving, with other property, a loss of about one hundred and thirty thousand dollars. Every hotel, every saloon, and every business stand of importance in the place was destroyed. Scott's brick house, containing the offices of the Eureka Canal Company and the office of Wells, Fargo & Co. on Main Street, escaped uninjured. (*Mountain Democrat,* August 8, 1856; *Sacramento Daily Union,* August 11, 1856)

Two years later, on August 31, 1857, Scott sold the Eureka Canal Company to Lewis B. Harris for $37,000. (Book C, Page 559) Harris was heavily involved in mining and ditch building in El Dorado County. He went on to become a partner in the South Fork Canal Company in 1865. (Book K, Page 543)

Sale of Bradley, Berdan to Harris

Soon thereafter, Harris bought Bradley, Berdan & Company, which, despite the loans from O.D. Mills, had been forced into bankruptcy. On September 9, 1857, in a sheriff's sale, El Dorado County Sheriff E.B. Carson sold Bradley, Berdan & Company to Harris for $20,000. (Starns 2004)

Bradley, Berdan Assets Added to Eureka Canal

Harris promptly added the assets of Bradley, Berdan & Company to the Eureka Canal Company.

After the addition of the Bradley, Berdan assets in 1857, the Eureka Canal Company had 247 miles of ditches and laterals, which started at Sly Park on the Cosumnes River and extended to within seventeen miles of Sacramento. (*Mountain Democrat, January 19, 1878*)

New Eureka Canal Company Incorporated

Harris changed the character of the Eureka Canal Company to a corporation in 1858. The Eureka Canal Company filed articles of incorporation with the Secretary of State's office

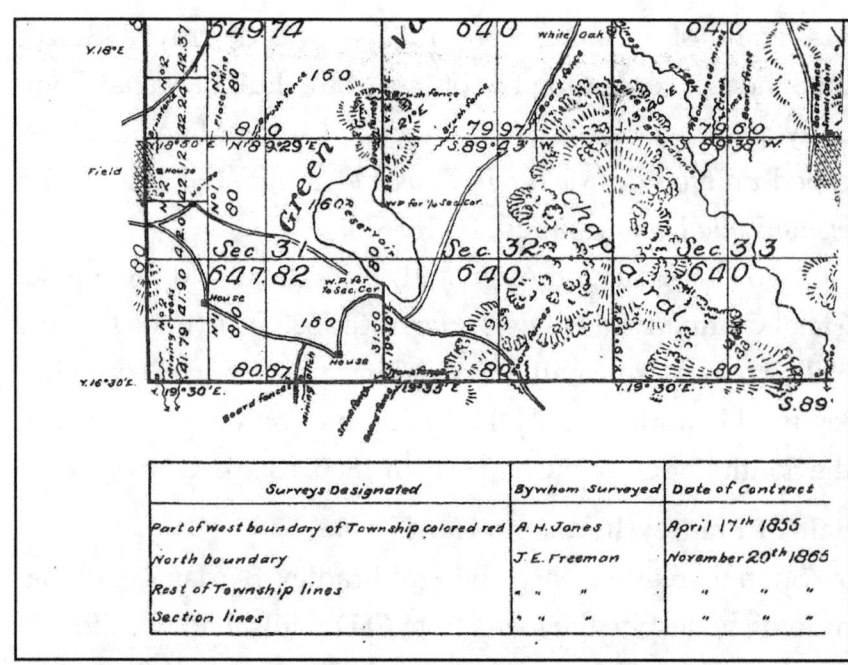

A detailed view of the 1865 White Oak Township survey that shows the American Reservoir in Sections 31 and 32

in Sacramento on January 13, 1858. (*Sacramento Daily Union,* January 14, 1858)

Scott Named Eureka Canal General Superintendent

Several month later, at a meeting of the Eureka Canal Company held in Sacramento on April 13, 1858, Lewis B. Harris, J. M. B. Wetherwax, John Bender, F. Tukey, and A.K.P. Harmon were elected Trustees (corporate directors). Harris was elected President; J. M. B. Wetherwax, Secretary; and W. P. Scott, General Superintendent. (*Sacramento Daily Union,* April 15, 1858)

Scott Irks Eureka Canal Board

At the regular August 1858 board meeting held in Diamond Springs, the corporation's trustees fired General Superintendent Scott by a unanimous vote. No details were disclosed as to what prompted the trustees to discharge Scott. At the urging of Harris, Scott was allowed to remain for three months to give him an opportunity to resign, but Scott was left without any authority to incur any debts on behalf of the Eureka Canal Company.

At the end of his three month grace, Harris, contrary to the wishes of the Board of Trustees, had not yet discharged Scott. Understandably upset, the **Board in November published a notice** to all "miners, water agents, ditch tenders, laborers and lumber men, and all other persons furnishing labor or materials" to the Eureka Canal Company, warning them that all contracts made by Scott on behalf of the Eureka Canal Company were illegal, not binding on the Company, and would not be paid.

It was disclosed that the trustees were upset because Scott had spent an estimated $25,000 in extending the Eureka Canal to the American Reservoir, at or near Clarksville, by enlarging the Buckeye race and fluming over Buckeye Flat by way of Shingle Springs House and Duroc House. (*Sacramento Daily Union,* November 23, 1858)

No doubt Scott was responding to the needs of the miners in the southwest portion of the county and saw a business opportunity in furnishing water to the region. All that portion of western El Dorado county extending from Weber Creek on the north to the Cosumnes River abounded in good diggings, both surface and ravine. But west of Buckeye Flat the area was un-watered, except by the rains of winter.

Miners' Water Needs

The tributaries of Carson Creek and Deer Creek had been worked every winter since the fall of 1850 with rocker and tom, and had paid and were yet paying good wages when a wet season afforded a sufficiency of water. After 1855, however, the best pay was in the flats and heads of ravines where water could not be had from the rains.

In 1856 the miners of this section endeavored to organize a company with the view to introduce water from Weber Creek. The miners of Jay Hawk, Carson Creek and Western Diggings, Plunkett's Diggings, and Marble Valley offered to work during the summer on any line of ditch that would bring water into their mines. Ninety of the miners offered to board and find themselves work at reasonable wages and take their pay in water when it came, provided the ditch company agreed to give security that a good supply of water would be furnished promptly upon the completion of the ditch.

The miners' attempt was ultimately abandoned. Though hopes were held out and promises made by both the Eureka and South Fork Canal Companies, promises which led the miners to believe that they would obtain water from these ditches at an early date, nothing came of this effort. A few surveys were made, and some ground was excavated on the proposed line of ditch, but in February of 1857 the ditch stood incomplete and unproductive. (*Sacramento Daily Union*, February 15, 1857)

~ 8 ~

CONSTRUCTION OF AMERICAN RESERVOIR (BASS LAKE)

In 1858, under Scott's management, the Eureka Canal Company extended its ditch system westward to the American Reservoir (now Bass Lake) (Peabody 1988).

When it was originally constructed, the American Reservoir was designed to impound water for the mining of placer gold. Dug by hand, the reservoir's capacity was approximately 500 acre-feet. Two pillars made of brick from Marble Valley were built by Chinese miners, and from these pillars an outlet gate was suspended. With this structure the gate could be raised and allow water out of the dam through the handmade conduit and into Carson Creek, which ran to the south through Clarksville.

It is not clear whether the American Reservoir was constructed as part of Scott's enlargement and extension of the Eureka Canal. However, this is the first documented reference to the American Reservoir and shows that the reservoir was there in 1858. Many years later American Reservoir would be renamed Bass Lake.

Evidence of Reservoir on GLO Plat

An 1865 General Land Office Plat of White Oak Township indicates the American Reservoir, labeled "RESERVOIR" on

54 *Bass Lake: A Gold Rush Artifact*

A poor photograph of Bass Lake when it was dry in 1978, showing the contour of the lake. Note the brick water-gate in the left of the picture.

*Photograph of the brick water-gate at Bass Lake,
believed to have been taken in 1978 when the lake was dry.*

the map. This is the first time we have found the reservoir documented on a map. Other landmarks on the same map include William Rust's Pleasant Grove House on Green Valley Road, which appears due north of the reservoir. Rust is buried in the little pioneer cemetery across from Pleasant Grove House, on the south side of Green Valley Road.

Since water flowed continually into the American Reservoir from the Eureka Ditch, it follows that some way of releasing the water was necessary, or the reservoir would simply fill up and overflow. We will see later that the town of Clarksville obtained a supply of water from the reservoir by way of Carson Creek, which passes closely to the west of the reservoir and flows in a southerly direction through the town. The Environmental Impact Report for the El Dorado Hills Specific Plan notes that "[t]he north branch of Carson Creek...drains Bass Lake." No doubt a lateral ditch was built from the reservoir to Carson Creek to handle the water outflow, and the reservoir outflow augmented Carson Creek, especially in the dry summer months.

By 1868, El Dorado County had 833 miles of mining ditches, dominated by the Eureka Canal system, which operated 450 miles of ditches. In contrast, Placer County had 699.5 miles, and Trinity County had 139 miles, of mining ditches (Browne 1868, 203).

~ 9 ~

HORACE GREELEY ON MINING CANALS

The legendary newspaper editor Horace Greeley was one of the most influential Americans of the 1800s. He founded and operated the *New York Tribune*, one of the best and most prominent newspapers of the period.

Greeley's Advice: Go West

Greeley was not an ardent expansionist, but he enthusiastically supported an orderly westward movement. The admonition "Go West, young man," is often attributed to him.

Horace Greeley

What he did say in 1846 was, "The best business you can go into you will find on your father's farm or in his workshop. If you have no family or friends to aid you, and no prospect open to you there, turn your face to the great West and there build up your home and fortune." This is often shortened and paraphrased to, "Go West, young man, and grow up with the country."

Greeley Goes West

In 1859, Greeley followed his own advice and went west to gather information that would interest the readers of the *New York Tribune*. The letters he wrote back to the *New York Tribune* during his journey through Kansas, Utah, and California were published on the Tribune in 1860. The following exerpts are taken from his letter from Sacramento on August 7, 1859:

> I have spent the last week mainly among the mines and miners of El Dorado, Placer, and Nevada counties, in the heart of the gold-producing region. There may be richer 'diggings' north or south; but I believe no other three counties lying together have yielded in the aggregate, or are now producing, so much gold as those I have named. Of course, I have not been within sight of more than a fraction of the mines or *placers* of these counties, while I have not carefully studied even one of them; and yet the little information I have been able to glean, in the intervals of traveling, friendly greeting, and occasional speech-making, may have some value for those whose ignorance on the subject is yet more dense than mine.
>
> The rivers come brawling down from the Sierra Nevada through very deep, narrow valleys or cañons,

... with a good portion of those of the rivers being mainly drawn off into canals or 'ditches,' as they are inaccurately termed, by which the needful fluid is supplied to the miners.

These canals are a striking characteristic of the entire mining region. As you traverse a wild and broken district, perhaps miles from any human habitation or sign of present husbandry, they intersect your dusty, indifferent road, or are carried in flumes supported by a frame-work of timber twenty to sixty feet over your head.

Some of these flumes or open aqueducts are carried across valleys each a mile or more in width; I have seen two of them thus crossing side by side. The canals range from ten to sixty or eighty miles in length, and are filled by damming the streams wherefrom they are severally fed, and taking out their water in a wide trench, which runs along the side of one bank, gradually gaining comparative altitude as the stream by its side falls lower and lower in its cañon, until it is at length on the crest of the headland or mountain promontory which projects into the plain, and may be conducted down either side of it in any direction deemed desirable.

Several of these canals have cost nearly or quite half a million dollars each, having been enlarged and improved from year to year, as circumstances dictated and means could be obtained. One of them, originally constructed in defiance of sanguine prophecies of failure, returned to its owners the entire cost of its construction within three months from the date of its completion. Then it was found necessary to enlarge and every way improve it, and every dollar of its net earnings for the next four years was devoted to its perfection. In some instances, the projectors exhausted their own means and then resorted to borrowing on mortgage at California rates of interest; I learn without surprise that nearly or quite every such experiment resulted in absolute bankruptcy and a change of owners. Of late, the solvent and prosperous companies have turned their attention to damming the outlets of the little lakes which fill the hollows of the Sierra, in order to hold back the superabundant waters of the spring months for use in summer and autumn. This course is doubly beneficent, in that it diminishes the danger from floods to which this city is specially subject, but which is also serious in all the valleys or cañons of the mining region wherein there is anything that water can injure.

I judge that the cost and present cash value of these mining canals throughout California must be many millions of dollars, paying in the average a fair income, while their supply of water is at this season, and from July to November, utterly inadequate.

Water is sold by them by the cubic inch—a stream four inches deep and six wide, for instance, being twenty-four inches, for which, at fifty cents per inch, twelve dollars per day must be paid by the taker. A head of six inches—that is, six inches' depth of water in the flume above the top of the aperture through which the water escapes into the miners' private ditch or flume—is usually allowed.

The price per day ranges from twenty cents to a dollar per inch, though I think it now seldom reaches the higher figure, which was once common. Were the supply twice as copious as it is, I presume it would all be required; if the price were somewhat lowered by the increase, I am sure it would be. Many works are now standing idle solely for want of water.

~ 10 ~

EUREKA CANAL COMPANY UNDER D.O. MILLS

In 1873 the Eureka Canal Company was purchased by Darious Ogden Mills, Edgar Mills, and their partner Henry Miller. At that time the company assets consisted of the old Bradley and Berdan ditch system, combined with the Crawford Ditch system consisting of the old Jones and Furman ditches, the North Fork Extension, the High Camp Ditch, and other smaller ditches such as the Weber Creek Ditch, later known as the Farmer's Free Ditch. The High Camp Ditch carried water from the new diversion at Baltic Mill through Sly Park and down to the area of Fort Jim and Newtown, then on through Diamond Springs, El Dorado, and then to the American Reservoir. From this reservoir ditches carried water on to Clarksville and surrounding area mines and ranches.

Darius Ogden Mills, Banker

Darius Ogden Mills was a prominent American banker and philanthropist. For a time, he was California's wealthiest citizen. He was born in North Salem, New York, on September 25, 1825, and in his early career he was a bank clerk and retailer. He joined the California Gold Rush in December 1848.

Soon tiring of mining, Mills opened a mercantile establishment in Sacramento. He began storing gold for the miners,

Darius Ogden (D.O.) Mills

and later began buying gold and issuing notes that circulated as money. Within a few years he changed the sign on his building from "store" to "bank." Mills founded the National Bank of D. O. Mills around 1849. A branch bank was opened in Columbia, Tuolumne County, California in 1850, which was transferred to other banking interests in 1859. Mills never invested in gold mining or silver mining directly, as he considered mining to be too speculative. Instead, he fostered ancillary businesses that supported the mining industry, such as banks, railroads and ditch companies.

Bank of California

In 1864, with other investors, Mills founded the Bank of California, with Mills as president, and William C. Ralston, Sr. as chief cashier.

William C. Ralston Sr. was born January 12, 1826, in Ohio. Ralston settled in San Francisco in 1854 and spent the next ten years with various banking firms. As chief cashier of the Bank of California, he invested much of the young bank's money into the silver mines of Nevada's Comstock Lode and as a result, Mills, Ralston, and the bank profited greatly in the 1860s and 1870s. However, crippled by Ralston's overspending and failed ventures, the Bank of California collapsed on August 26, 1875, and Ralston was asked to resign. His drowning in San Francisco Bay on August 27, 1875, was ruled an accident.

Mills used his personal fortune to revive the Bank of California. The bank reopened on October 2, 1875, with $2 million in gold coin on hand, and Mills again as president.

Millbrae

In the 1860s Mills purchased 1,500 acres of land west of what is now the San Francisco Airport and there built a 44,000

Millbrae Estate

square-foot mansion on the 1,500-acre property, which he called Millbrae.

The Mills' Millbrae mansion remained intact from the mid-1860s until the mid-1950s, giving its name to the town that grew up around it. By the end of the 1930s, the aging Millbrae mansion had outlived its usefulness as a vacation home for the extended Mills family, and it fell into disrepair.

The property was sold to developers after the mansion was destroyed by fire in 1954.

In 1872 the **National Bank of D. O. Mills** became the National Gold Bank of D. O. Mills & Company. It remained under that name until October 29, 1925, when it was merged with the California National Bank of Sacramento.

Later in life, when Mills retired from banking, he returned to New York and turned to philanthropy. When he died on January 3, 1910, he left an an estate worth more than $36 million.

~ 11 ~

PARK CANAL & MINING COMPANY

Mills and his partners kept the Eureka Ditch Company for only two years. Though improvements to the ditch system were planned, they were never completed. Mills became concerned about the expenditures of his banking partner William Ralston, and Mills would undoubtedly be aware of the problems associated with owning and maintaining large ditch systems, in spite of the need for such systems. (Starns)

Eureka Ditch Purchase

In 1875 the entire Eureka Ditch Company property was purchased by J. M. Crawford and others, of Philadelphia, under the title of Park Canal & Mining Company (Limited), incorporated on November 11, 1875, in Philadelphia, Pennsylvania. The officers of the company were J.M. Crawford, chairman; Samuel F. Fisher, Secretary and Treasurer; J.J. (John Jones) Crawford (J.M. Crawford's brother), General Manager; and M. G. Griffith and Samuel Hale, Superintendents. The corporation's principal office was 308 Walnut Street, Philadelphia, with branch offices at Diamond Springs and Dry Gulch in the gold country. (Sioli; Pennsylvania Secretary of State)

Under J.J. Crawford, the Park Canal and Mining Company flourished. A popular and admired personage, Crawford is credited with energetically enlarging and improving the canals and ditches, so much so that the main ditch of the Park Canal and Mining Company became known as Crawford's Ditch or Crawford Ditch. The following life of J.J. Crawford is based for the most part on works by local historian George Peabody.

J.J. Crawford

Crawford was born in New Castle, Pennsylvania on February 12, 1846, one of two sons of John McFarland and Elizabeth Jones.

Crawford was reportedly an attentive student, and his parents encouraged his studies in the fields of mining and engineering. During his late teens, he served in the Union Army in the Civil War. Despite that interruption, he graduated from the Polytechnic College of Pennsylvania in 1867 when he was twenty-one. Then Crawford traveled to Europe, where he attended the Royal School of Mines in Freiberg, Germany, graduating in June of 1870.

John Jones (J.J.) Crawford

Back to the United States from Germany, he traveled to Nevada where in 1871 he became superintendent of the Great Basin Mining Company in White Pine County at age twenty-five.

Crawford left a well-paying job in Nevada to accept the position of General Manager of the Park Canal & Mining Company. He took up residence in the now-historic Gay family house in Dry Gulch on the north side of Sly Park Road near the intersection of Twin Cedars Road.

Park Canal Operations

Crawford established offices in Dry Gulch and in Diamond Springs. Then, with Pleasant Valley's lumberman and ditch genius E. V. Davenport, work was commenced rebuilding and enlarging the 250 miles of old miners' ditches. They added fifty miles of new ditches, extending them from the Steeley Fork of the Cosumnes River (near Cole's Station), from Camp Creek far above Sly Park, all the way west to within 17 miles of Sacramento County. Those ditches, enlarged and extended, brought greater supplies of water for use in his Dry Gulch hydraulic mine, and to Pleasant Valley, to Newtown, Fort Jim, Ringgold, Hanks Exchange, Diamond Springs, Missouri Flat, El Dorado, Shingle Springs and westward to the American Reservoir (today's Bass Lake near Cameron Park).

Being a mining engineer, Crawford developed the huge Dry Gulch Mine for hydraulic mining. Today you can see its great scar in the earth just south of Sly Park Road below Meadow Lark Way. The clean-up of his sluice boxes after his first winter's experimental run supplied him with $35,000 in gold.

Crawford assisted C. P. Steadman in developing the profitable Dead Head hydraulic mine on Dead Head Gulch on the west side of Newtown, using water from the Park Canal via the Steadman ditch which ran along the northern slopes of Newtown Ridge. Of course, Crawford's Park Canal was a source of income for the ditch company, too, and he encouraged the use of its waters that were surplus to his needs. The Park Canal, especially the Crawford Ditch from Clear Creek, Camp Creek and the Steeley Fork of the Cosumnes River, provided water for the development of other mines, for irrigating and expanding existing orchards and vineyards, and for lumber operations.

This abundant water supply resulted in the El Dorado Fruit Farm of Hanks Exchange becoming the largest fruit ranch in El Dorado County; supplied water power to operate the machinery of one of California's great gold producers, the underground Grand Victory Mine of Hanks Exchange; provided the log ponds for the giant California Door Company lumber mill in Diamond Springs; and furnished the water to operate the Westinghouse electric generator in Ladies Valley. In addition, reused water from the Park Canal dropped into seasonal creeks such as Squaw Hollow Creek, Ringgold Creek, Tiger Lily Creek and Martinez Creek, increasing their flow to the delight of the miners and sustaining the riparian wildlife. It was Crawford who envisioned major water storage facilities where Jenkinson Lake lies today.

J.J. Crawford Family

In 1880, Crawford married Fannie Morey (*Mountain Democrat,* January 10, 1914). Her brother, Henry S. Morey, was a machinist and an influential entrepreneur in the Sly Park and Placerville areas, and he later owned the Placerville Foundry.

In 1889, Henry Morey had been the highest bidder at a Sheriff's sale for the High Camp Creek Ditch engineered by his brother-in-law (El Dorado County Sheriff's Certificate of Sale, Book B:78,80).

Henry Morey also purchased (for $1,700 as lowest bidder) the Park Creek Ditch which began at a point on Park Creek about one half mile below or west of Sly Park. It extended five and one-half miles westerly to the Dry Gulch Gravel Mine (El Dorado County Sheriff's Certificate of Sale, Book B:80) (Map Thirty). The mine contained a water powered derrick that Morey obtained along with a tract of land and the American Reservoir (today's Bass Lake in El Dorado Hills). Morey sold the land to John and James Blair in 1893 (El Dorado County Deeds, Book 43:563).

Fannie Crawford was said to be very dedicated to her husband and children. However, in 1888, after giving Crawford two daughters, Daisy and Bertha, Mrs. Crawford fell ill with typhoid fever. Crawford quickly packed up his family and took them back to his family home in Philadelphia, where Mrs. Crawford and the children would be cared for while she was tended by the best doctors in the country. Crawford returned to California to manage the family business on Diamond Ridge, selling all their household goods and finding room and board.

Though concerned about his family in Philadelphia, Crawford continued in his business of superintending the Park Canal & Mining Company. He also expanded his interest in mines and canals in Placer County, Amador County and in Calaveras County. Crawford was drawn to public welfare and civic activities, becoming El Dorado County's Anti-Chinese representative, an effort to protect jobs for United States citizens locally.

During the year 1890, Mrs. Crawford recovered sufficiently to return to Placerville with the children. Crawford had purchased a home in town near schools, churches, and the active social activity where his family would prosper.

J.J. Crawford Public Life

That same year, Governor Robert W. Waterman appointed Crawford a member of the Examining Commission on Rivers and Harbors, which was concerned with navigational hazards resulting from miners hydraulicking for gold in the foothills. He served unselfishly on the Commission, armed with his knowledge and experience as a Mining Engineer, though to his own personal disadvantage as the operator of the Dry Gulch hydraulic mine. Commissioner Crawford was a regular delegate to the State Miners' Association, and he was a member of the American Institute of Mining Engineers.

Crawford often took his chidren with him on business trips so that they might become acquainted with California's cities and with his affairs of business. He was appointed Director of the Placerville Schools and was elected to the Board of Education. Crawford was an active officer in the Knights of Pithias, the Secretary of the Masonic Hall Association, a political leader in the Republican County Conventions, and a member of the Grand Army of the Republic.

At age 47, Crawford was appointed by Republican Governor Henry H. Markham to the post of California State Mineralogist, in which office he served from 1893 to 1897. While State Mineralogist, he published bulletins on topics such as gold mill practices and gas yielding formations. He also published a bibliography of California geology and a catalog of California fossils.

His work as State Mineralogist required that he move his family to San Francisco. Later the family relocated to the town of Alameda. However, El Dorado County remained home to the family. Crawford regularly returned to Placerville to participate in political affairs and civic activities, often bringing

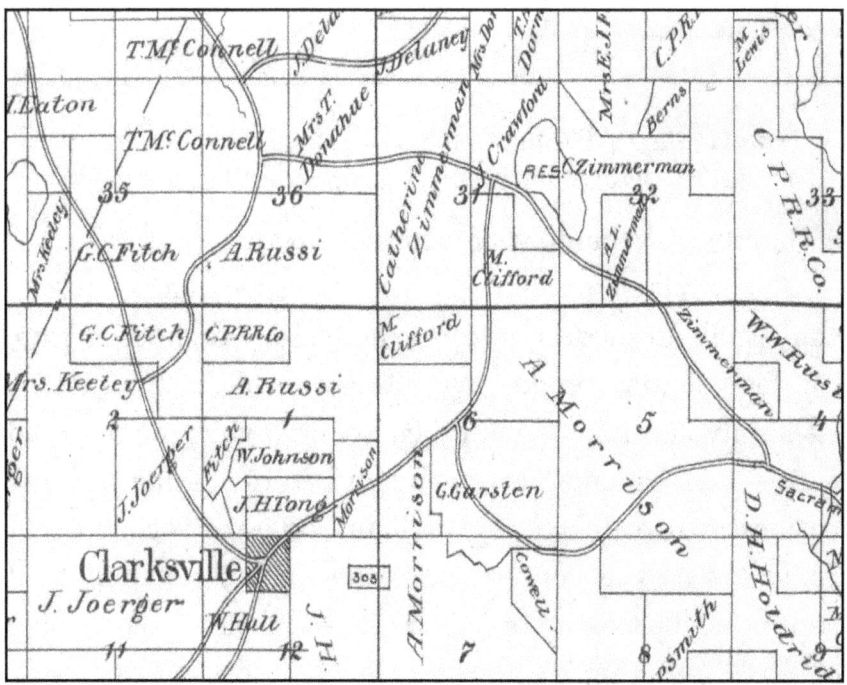

This detail of the 1895 Punnett map shows the American Reservoir ("RES") northeast of the town of Clarksville. J. Crawford is identified as the owner of the property on which the reservoir is located.

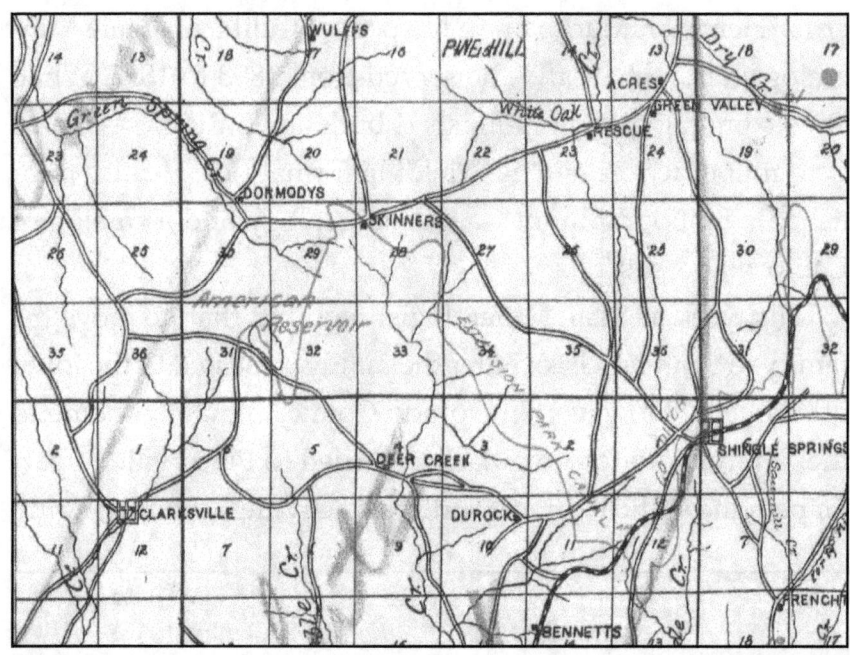

Detail of the 1909 California State Mining map showing the American Reservoir and the Park Canal extension which fed it.

his family with him to visit their old friends.

Historical references to Crawford describe him as a gentleman, well-educated, and a caring and family-oriented person. He was said to be progressive yet responsible, and trusted by his family and friends with their wealth, which he returned with good dividend. He was helpful and inspiring to small businessmen, active in civic, philanthropic and social institutions, and he was concerned for the welfare of the State of California, the County of El Dorado, and their citizens.

Park Canal Improvements

In 1877 under Crawford's leadership, the Park Canal & Mining Company built a substantial ditch, capable of carrying 1,800 inches of water—the old Eureka Ditch carried but

1,200—from Camp Creek, under the New Baltic mill, across Diamond Creek and Stonebreaker Creek, before dropping into Sly Park Creek in Hazel Valley. The system of the company's canals was such, that water used for mining purposes could be used again and again. It was available for distribution over a large area of country, and particularly adapted to the cultivation of vines, fruit trees and vegetables. The whole extent of ditches owned and controlled by the Park Canal & Mining Company became nearly 300 miles.

American Reservoir and Clarksville

Water from the American Reservoir (now called Bass Lake) is discharged through the watergate in the dam into nearby Carson Creek. This creek flows in a southwesterly direction through Clarksville and eventually joins the Cosumnes River at what is now Sloughhouse in Sacramento County. The reservoir-fed creek was an important water supply for Clarksville.

In 1887, the *Mountain Democrat* issue of August 27 published a story in the column "Around the Hub, The Latest News from Outside Districts" that covered the Clarksville district. Special Correspondent Mark Lee wrote: "Heretofore we have been able to procure a sufficient supply of water from the American Reservoir for all necessary irrigating purposes—that is for orchards and small gardens—which gave life and tone to our little homes and surroundings, but this year the great gate of that artificial lake has been closed and sealed up."

Park Canal and American Reservoir 1895 to 1910

A map of El Dorado County was drawn in 1895 by Pun-

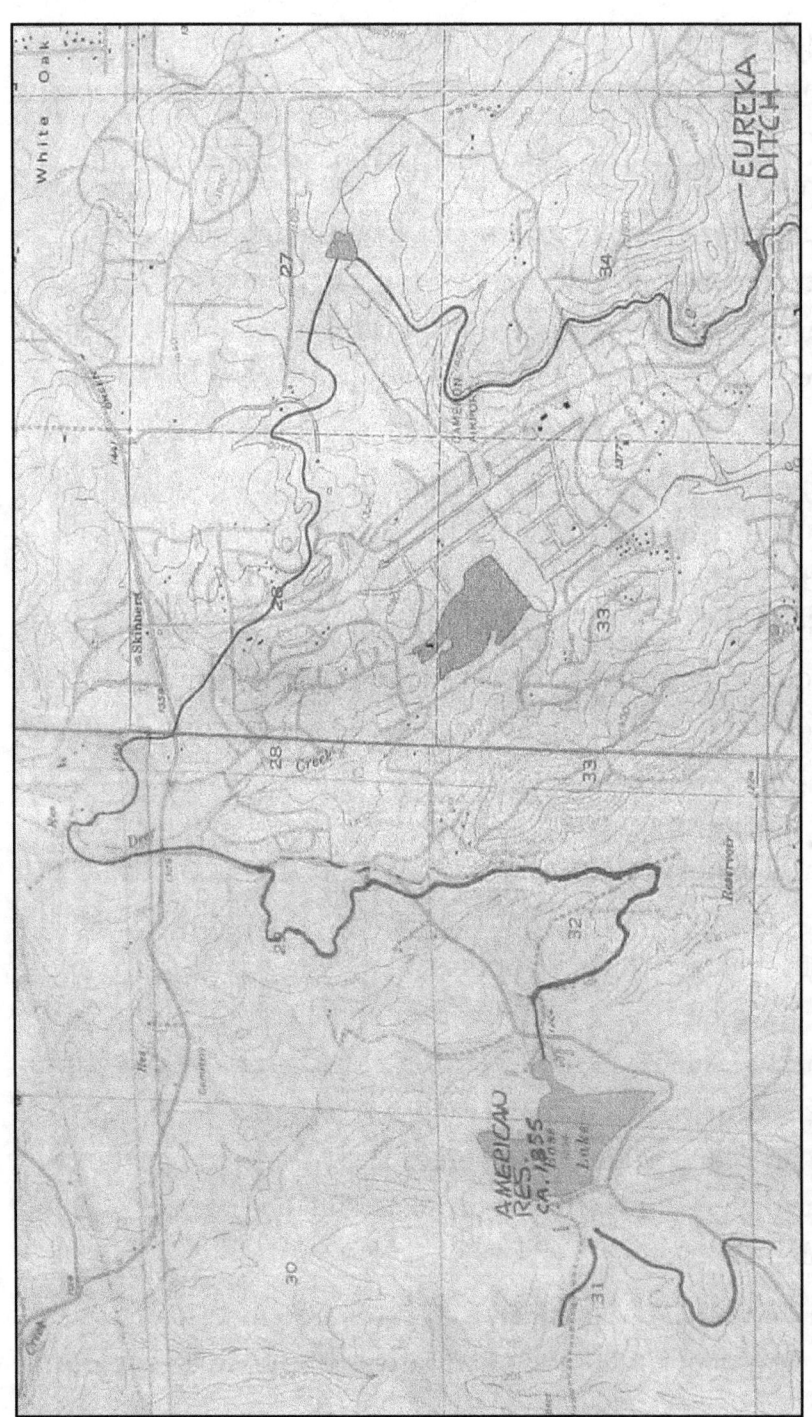

The approximate route of the Park Canal to the American Reservoir (EDC Historical Museum)

nett Bros. and said to have been "compiled from the official records and surveys" for Shelley Inch, bookseller and stationer, Placerville, California. On that map American Reservoir is shown as being on the property of one "J. Crawford."

In 1903, the *Mountain Democrat* listed a number of County Statistics compiled by the County Assessor. Among the lists was one that enumerated the mining and irrigation ditches. The Park Canal was credited with a length of 30 miles and an assessed value of $14,000.

At some time around 1909, the California State Mining Bureau issued a map of El Dorado County, California, showing the mining and water ditch activities located in the county. On this map the present Bass Lake is labeled "American Reservoir" and is shown as the terminus of an extension of what is labeled "Extension Park Canal and Mining Co. Ditch." The numbered claim or mine location markings written on the map have no legend.

Sale of Park Canal

In 1911, Crawford sold his water holdings, including the American Reservoir. The front page of the *Mountain Democrat* of May 9, 1911, carried the following article: "J. J. Crawford has sold the water rights and franchises commonly known as the Park Canal property, including the Eureka Canal system, the Camp Creek ditch, the Park Creek ditch, the American reservoir and the Diamond Ridge ditches, to Judge C. E. McLaughlin of Sacramento, as trustee for some undisclosed purchasers. The price is not given out, but a mortgage for $25,000 has been taken back by the seller. It is probable that the water will be used for irrigation of certain tracts of land in the neighborhood recently acquired by Sacramento parties."

Based on the documentation of the later 1917 transfer of the same property to Diamond Ridge Water Company, the undisclosed purchasers may be reasonably assumed to have been O. Scribner and Florence Ives Scribner (his wife) and Charles B. Bills and Ella C. Bills (his wife).

~ 12 ~

DIAMOND RIDGE
WATER COMPANY

Evidence indicates that between 1911 and 1917 the Park Canal system, or the Crawford Ditch system as it was popularly known, continued to supply its mining and farming customers with water under the management of Judge McLaughlin, as trustee, and the Scribners and the Bills.

A description of the assets owned by this group appeared in the 1913 delinquent property tax list of the May 14, 1913, *Mountain Democrat*.

There was displayed a notice that the County Assessor had levied "Against C. E. McLaughlin and the franchise of the ditch constructed by Park Canal and Mining Co. Limited, from Camp and Park creeks known as the high lines, with water rights, also franchise of mining and irrigating ditches including syphon, pipe, flumes, etc., formerly known as the Eureka Canal formed by the consolidation of the two systems and ditches known as the Jones Firman & Co. and the Bradley, Berdan Co. All the foregoing conveying waters from the North and Steely Forks of Cosumnes River; also from Camp, Park, Clear, Niger Lily [sic], Ringgold, Slate and Deer creeks and including the right to the waters in said streams, extending to the Dry Gulch mine, Pleasant Valley, Newtown, Fort Jim, Diamond Springs, El Dorado, Arum, Shingle Springs,

Clarksville and other places to the western limits of El Dorado county, 30 miles at $466 taxes, penalties and costs." This description certainly included the American Reservoir.

Formation of Diamond Ridge Water Co.

The Scribners and the Bills formed the Diamond Ridge Water Company, which was incorporated on March 23, 1916.

The *Report of the California Railroad Commission* (1922) lists the company's business address as 58 Sutter Street in San Francisco. The officers are listed as O. Scribner, President and General Manager; Chas. B. Bills, Vice President; and M.B. Downing, Secretary and Treasurer.

Transfer of Eureka Canal to Diamond Ridge

In 1917, six years after purchasing the ditch system from Crawford, the group made up of C. E. McLaughlin, Trustee, O. Scribner and Florence Ives Scribner (his wife) and Charles B. Bills and Ella C. Bills (his wife), deeded "all that system of canals, ditches, races, flumes, rights of way, pipes, reservoirs, dams, water rights and franchises described in Deeds, Book Q:12, known as the Eureka Canal Company, Jones, Bradley and American Reservoir extension main trunk canals or ditches" to the Diamond Ridge Water Company. (El Dorado County Deeds, Book 96:444). This deed certainly included the American Reservoir.

That the reservoir was included is evidenced by the fact that some eight years later the Diamond Ridge Water Company was listed in the *Mountain Democrat* of June 12, 1925, as being delinquent in its property taxes for the year 1924. The notice listed all of the properties so assessed, and included among the list: "(1) all that system of canals, ditches, races, flumes, rights of way, pipes, reservoirs, dams, water rights, and franchise of the Eureka Canal Co. of El Dorado, known and described as Jones, Bradley and American Reservoir extension. . . . (6) High Camp Creek and Newtown main line from Baltic or Camp Creek to Diamond Springs via Newtown, Diamond Springs, American Reservoir main line. . . . (14) Two Diamond Springs Reservoirs (also American Reservoir tract)."

~ 13 ~

EL DORADO COUNTY WATER USERS ASSOCIATION

The expansion of Western States Gas and Electric into hydro-electric power production began to concern local farmers regarding water rights on the South Fork of the American River.

Western States Expands Water Use

In 1916, Western States enlarged the upper portion of the Main Ditch and constructed Forebay Reservoir near the South Fork of the American River. Water was diverted to their American River powerhouse, which was known as the El Dorado Power Plant. That diverted water was seen by the Camino and Placerville farmers as no longer being available for irrigation.

Railroad Commision Complaint Filed

The intentions of Western States impelled the water users on the canal system to form the El Dorado County Water Users Association and to file a complaint before the California Railroad Commission (predecessor of the Public Utilities Commission).

The Water Users Association asked the Railroad Commission to determine the obligations of the electric company to the agricultural water users. The Commission's decision in the matter held that all water controlled by Western States had been devoted to public use and that there is no preference on the question of public use between irrigation and hydro-electric use; that the obligation of the utility was not limited merely to the water delivered to past consumers but that it could be required to make reasonable additions to its system to provide new service; that the utility could devote water to hydro-electric purposes when service to existing consumers is provided for.

Western States Concedes Canal Assets

To avoid litigation and the need to furnish water to others, Western States sold that portion of the El Dorado ditch system below the Fourteen Mile House tunnel to the El Dorado Water Company, a company formed by the Water Users Association, for $25,000 in bonds of the new company. Western States also agreed to deliver to the El Dorado Water Company 40 second-feet, or 1600 miner's inches of water, for irrigation use, and five second-feet, or 200 miner's inches, for mining use.

The sale and service agreements were authorized by the Railroad Commission in its Decision No. 6436, dated June 25, 1919. The electric company then sold water to the El Dorado Water Company on a wholesale basis according to the terms of the service agreement, and the water company became the only consumer being served with water from the El Dorado ditch.

The Water Users Association was the organization out of which the El Dorado Irrigation District would emerge, with its interest and ownership in the Crawford Ditch system, the Main Ditch, and associated ditches on the Placerville Divide.

~ 14 ~

El DORADO IRRIGATION DISTRICT (1925-1955)

The demand for irrigation and domestic water continued to increase, and it soon became apparent that this amount of water was not sufficient to meet the increasing demand. It was, therefore, decided to develop a water supply on Webber Creek by constructing a concrete dam and about six miles of ditch to connect with the Water Company's ditch at Placerville.

El Dorado Water Company Reorganization

To facilitate the financing of the project, the El Dorado Water Company was reorganized under the same ownership and management. This resulted in the incorporation of the El Dorado Water Corporation in February 1922. Necessary water rights were secured, and construction of Webber Dam and conduit was started. There was still a shortage of water, however, and in February 1923, the Corporation obtained an option to purchase the rights and property of the Diamond Ridge Water Company, paying $10,000 on a total price of $50,000.

This original payment was later forfeited and the project abandoned. Financial troubles beset the Water Corporation and prevented the full completion of the Webber Dam Project. They also brought about the abandonment of the development, which had been planned to meet the ever growing needs of the community. El Dorado Water Corporation was in trouble.

Enter the El Dorado Irrigation District.

El Dorado Irrigation District

The formation of the El Dorado Irrigation District was authorized by a General Election held in September 1925 on the question, "Shall El Dorado Irrigation District be Formed?" When the votes were canvassed by the Board of Supervisors on October 5, 1925, it was declared by that body that the electors had decided in favor of the proposition and that El Dorado Irrigation District "was duly formed." At the same time, El Dorado Irrigation District assumed the obligations and assets of the El Dorado Water Corporation.

The District, organized under the Irrigation District Law of the State of California, is composed of five divisions, comprising 31,560 acres of land. The City of Placerville is designated as one of these divisions. The members of the first Board of Directors were William A. Caldwell, Albert G. Volz, Nicholas Fox, George M. Smith, and Charles H. Clifton. Other elective officers are the Assessor and the Collector-Treasurer. The Secretary-Manager and the Attorney for the District are appointed by the Directors.

New Water Supply Sought

An interesting quote from the Minutes of the first Board of Directors' Meeting held at the office of the El Dorado Water Corporation at 9 Coloma Street, Placerville, California, on Oc-

EID Headquarters bulding on Main Street in Placerville, bought in 1927 and used until 1957 when EID moved to its present headquarters on Mosquito Road.

tober 5, 1925, reads: "The question arose as to the water shortage for the coming irrigation season, and it was the consensus that every effort be made to bring about immediate relief." The struggle with the water shortage problem was the very first that faced the newly-formed District twenty-seven years ago. As the El Dorado Water Corporation was also working to improve the water supply, joint meetings were held between that organization and the District so that there would be no duplication of work but rather a combined effort to achieve the desired result. The Directors of the Irrigation District employed an engineer to study the water problem and eventually decided, after studying the engineer's report, that the best solution would be to go to Hazel Valley or Sly Park and bring water into the District from those sources. Options were obtained for the purchase of property to make such a development possible.

In February 1927, a Bond Election was called in order to finance this development. The election carried, and a $1,300,000 bond issue was authorized. Of this amount, $685,000 in bonds were issued. On June 1, 1927, the El Dorado Irrigation District took over the El Dorado Water Corporation and assumed all its obligations and assets.

The transaction called for an expenditure of $365,527.63. The proposal at this time was for a storage dam to be built at Hazel Valley to supplement the present water supply. The new project was vigorously prosecuted. A lot was purchased in the City of Placerville, and the District's present office building located on Main Street was erected. The options that had been obtained to purchase lands in the Hazel Valley area that were needed for the project were exercised in the amount of some $80,000.

Prior to 1929, the Water Company furnished the City of Placerville with water. In 1929 the District built a filtration and chlorination plant at Sacramento Hill at a cost of $20,000 to purify the water for the City. The following year the District turned the plant over to the City for the sum of $1.00. Placerville has continued from that time to operate its own facilities.

Difficulties Encountered

The District ran into trouble, financial and otherwise, and the Hazel Valley project was eventually abandoned. The Directors decided to enlarge Webber Dam instead. More land was purchased at Webber Dam, and plans for a hydraulic earth-fill dam were completed. A contractor was found who made a proposal and bid, which were acceptable to the Directors, and a contract was drawn on June 12, 1930, proposing to build Webber Dam, a diversion tunnel, a waste way, and appurtenances.

After this work had been started, real trouble beset the District. Dissension, panic, hard times, bankruptcy of the contractor, and the failure of the Bonding Company prevented completion of all of the improvements. The Great Depression of the 1930s had set in, and the El Dorado Irrigation District was not the only organization to experience trouble at this time. Banks were failing throughout the country; pears, a major crop in El Dorado County, were bringing in "red ink;" and taxes were in default.

Federal Aid Sought

It became impossible for the District to meet its obligations, and all development work ceased. The Directors finally sought aid from the Government, and in 1933 a request was made to the Reconstruction Finance Corporation for a loan to refinance its $688,000 indebtedness. This request was refused on June 5, 1934. In August of this same year, a new request to the Reconstruction Finance Corporation was made for a loan, and eventually the Agency approved a loan of $360,500 at four per cent interest, provided that the Irrigation District would buy up all its outstanding bonds at $0.505 on the dollar, with all the coupons attached.

In March 1935, the California Districts Securities Commission approved the Directors' request for a refunding issue of bonds in the amount of $360,500. A Bond Election called to ratify this issue carried almost unanimously.

All development work on the Hazel Valley and Webber Dam projects had stopped, but there was still the urgent need for an additional supply of water. In 1938 the District availed itself of an opportunity to purchase the bankrupted Diamond Ridge Water Company for the sum of $3,232, plus the outstanding debts of the Company. The principal debt was an obligation to the California Door Company. This debt against the Diamond Ridge Water Company was cancelled when its purchase by El Dorado Irrigation District was consummated.

Ditch System in Need of Repairs

The ditch system of the Diamond Ridge Water Company was in a very sad state of repair, and the District was fortunate in securing from the Works Progress Administration rehabilitation aid in the form of labor amounting to some $80,000. Land was purchased at the head of Cedar Ravine Creek on the outskirts of Placerville, and the State of California estab-

Workers rehabilitating the Diamond Ridge Ditch during the Great Depression under a grant from the California State Relief Administration.

lished a camp under the State Relief Administration for some two hundred and fifty men.

After rehabilitation, the Diamond Ridge Ditch furnished a supplemental supply of water that almost took care of the Missouri Flat section of the District. However, there being no storage, the supply is still subject to the fluctuating stream flow of Camp Creek and the North Fork of the Cosumnes River.

Sly Park Project Favored

In 1938, Fred Hosking, Engineer-Manager for the District at that time, prepared surveys and plans for a new project at Sly Park. However, in 1939 the Districts Securities Commission refused permission for the District to proceed with this plan of development. Late in 1939 W. E. Jenkinson was appointed Manager to succeed Mr. Hosking, who passed away.

Four years later the Bureau of Reclamation was contacted relative to the desperate need for water in the District, and the steps necessary to receive Federal aid were outlined. In 1943 the possibilities of storing water on Squaw Creek to supplement the supply of water to Missouri Flat District were given very serious consideration. Surveys were made, and the feasibility of this plan was established. During that same year the demand for water was increased by the many people who planted victory gardens. There was simply not enough water to go around, and the shortage was extremely acute.

District Asks Study of Reclamation Bureau

Blakeley Reservoir was raised five feet, but even this additional storage was small compared to the greatly increased demand. So in 1944, the District advanced $5,000 to help defray expenses and called upon the United States Bureau of Reclamation to make a complete study of its water problems and to advise the District on the merits of the Squaw Creek and Sly Park projects. The Bureau was very cooperative and sent its geologists and engineers to look over both projects. They advised the District that Sly Park, although more expensive, would be the more desirable project to pursue.

A meeting was held in the office of the District in December 1945 between the Board of Directors and high ranking officers of the Bureau. At this meeting a contract was entered into whereby the Bureau was to make a preliminary engineering study of the Sly Park project, provided the District put up $5,000 of the estimated cost of $10,000. The Directors authorized the $5,000 expenditure, and the Bureau immediately put its engineers to work.

Beyond Ability to Repay

Further engineering studies, however, developed the fact that, although the plan was entirely feasible, it was probably beyond the financial ability of the District to repay.

Their problems were then laid before Congressman Clair Engle, who was quick to comprehend the District's critical situation. After many conferences with the Congressman, the question arose: "Why are we mountain people any different from our valley neighbors? They have their great Central Valley Project which is a widespread, co-ordinated development utilizing multiple purpose reservoirs. Its benefits are spread from Redding to Bakersfield, and the costs are reduced by the power generated by the falling waters—the water that originates in the mountain counties. The only difference appeared to be that the users of the Central Valley Project are below the large multiple reservoirs, while we in the foothills are above them."

Proposal Goes to Congress

At the very time the El Dorado Irrigation District was seeking a solution to the comparatively small but desperate needs of its mountain area, the Federal Government was working on plans to build Folsom Dam on the American River—a stream that derives a large part of its flow from El Dorado County.

The dam to be built would be in El Dorado County's front yard. It was obvious that no water from Folsom Reservoir could be diverted to the upstream land, but the people of El Dorado County could see no reason why they should not share in the power benefits.

Congressman Engle, with the backing of El Dorado County residents, took this proposal to Washington and made it a part of the American River legislation which placed Folsom Dam and the Sly Park Project in the Central Valley Project.

The resulting Bill H.R. 165 authorized the Army Engineers to construct Folsom Dam and the United States Bureau of Reclamation to construct the power house and Sly Park Dam. When the project was completed, the Bill authorized the Bureau to operate the dam and power house under Reclamation Law, and integrated the whole project with the Central Valley Project.

Bill Passed Over Stiff Opposition

The Bill met stiff opposition in Congress, and only the determined efforts of its proponents saved it from defeat. The Board of Directors, together with the assistance of a group of prominent men within the community headed by John W. Dunlop, took a very active part in securing the Bill's passage. Chronicled here, also, must be the untiring effort of Congressman Clair Engle in leading his Bill to its final enactment.

President Truman called the legislation the "Folsom Formula." Many people view it as merely a formula whereby a multiple purpose reservoir is constructed by one Federal Agency for transfer to another Federal Agency for operation as an integral part of an overall development.

A Deeper Meaning

The "Folsom Formula" had a much deeper meaning to the people of the El Dorado Irrigation District, as it marked a legislative accomplishment that could mean the solution of the water problems of mountain counties. To the El Dorado Irrigation District, it meant primarily that "upstream" water users were put on a par with "downstream" users insofar as the widespread financial benefits from a multiple purpose development by the Federal Government was concerned.

The District was then able to finance Sly Park, as a great portion of the cost was paid for from the power revenue derived from the electric energy sold from the Folsom Power House. The District continued to manage its own affairs, build its own distribution system and purchase water, canal side, from the United States Bureau of Reclamation.

Refinancing Saves Large Sum

In February 1946, the District had an opportunity to refinance its outstanding bonds, and in May 1946, called an election that resulted in practically unanimous approval of the refinancing plan. This refinancing saved the District some $33,000 in interest rates—3.75% as against the old rate of 4% paid to the Reconstruction Finance Corporation.

A small reservoir known as the Bray Reservoir was built at Diamond Springs in 1947. This was a regulating reservoir for the Missouri Flat Ditch and gave the District some storage at that point. At that time the District was getting back on its feet financially. Delinquent taxes, which stood at $6,000 a few years previous, were now less than $500. None of these delinquencies had to be foreclosed, and the District did not take title to a single acre or parcel of land.

Conflict Over Webber Dam Spillway

The Webber Dam was built under the jurisdiction of the Railroad Commission, not under the jurisdiction of the State Engineer's Office. This caused a conflict between the State Engineer's office and the District as to the size of the dam's spillway. Legislation passed after Webber Dam's construction provided that all dams in California above a certain storage capacity come under the supervision of the State Engineer's Office.

In the case of Webber Dam, the State Engineer's Office finally was able to compel the District to build a new spillway much larger than the original spillway, and to do other work which required expensive drilling.

We will return to the history of the El Dorado Irrigation District in Chapter 16.

~ 15 ~

BASS LAKE ROAD

Any narrative about Bass Lake, or the American Reservoir as it was once known, must include a mention of Bass Lake Road, which now passes close to, and to an extent loops around, the lake.

The present Bass Lake Road runs north from its intersection with Highway 50, swings around the south and east shores of Bass Lake, and then proceeds on to intersect with Green Valley Road near the Green Valley Cemetery. It behooves us to look at how things were in the 1850s in order to understand how the road came to be.

Early Road Activity

The first mention of the road that would eventually connect Highway 50 and Green Valley Road was in 1856. On September 2, 1856, the Board of Supervisors appointed Wm. Henry and Joseph Baxter as "viewers to view out and locate a road running from the Morrison House on the Sacramento and Placerville Road to intersect the Coloma and Sacramento Road at the Green Valley House." The road was to "commence at the said Morrison House and then run thence via of the Atlantic House and Big Reservoir [Bass Lake] to a point near the said Green Valley House."

The gold rush had fostered a number of wagon routes that led to the gold diggings. Many gold-seekers came by ship around Cape Horn to San Francisco, then up the Sacramento River to Sacramento, and finally overland to the gold fields. Travelers usually took one of two routes to the diggings: the Sacramento and Placerville Road, which eventually became Highway 50, or the Coloma and Sacramento Road, which eventually became Green Valley Road.

Early Road Houses

After gold was discovered and thousands flocked to the area, dozens of road houses were built along the routes between Sacramento and the gold fields in and around Placerville and Coloma. These road houses provided crude lodging

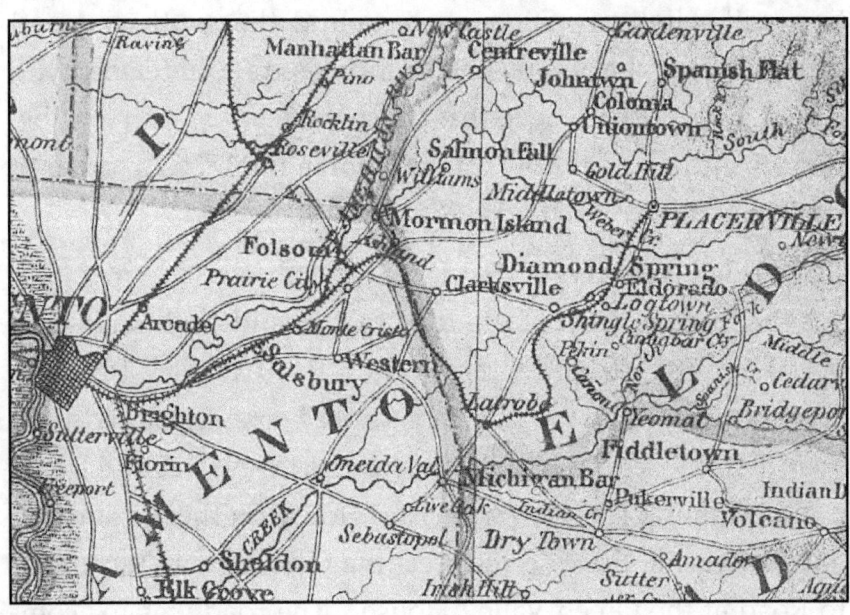

Detail of a California and Nevada map drawn by A. Craven, C.F. Hoffman and F. Leicht in 1874. The Sacramento-Placerville Road comes east from Folsom, passes through Clarksville and continues on to Shingle Springs and Eldorado. The Coloma Road forks off the Sacramento-Placerville Road and heads northeast through Uniontown to Coloma.

97 Bass Lake: A Gold Rush Artifact

and meals to teamsters and travelers. During the peak of the prospecting and mining period, there was a stopping place about every mile along every road to the diggings. These inns or "houses" were landmarks for early travelers.

The proposed road was to "commence at the said Morrison House and then run thence via of the Atlantic House and Big Reservoir [Bass Lake] to a point near the said Green Valley House." Let us see where these houses were, and what bearing they had on the north-south road that would one day connect Highway 50 and Green Valley Road.

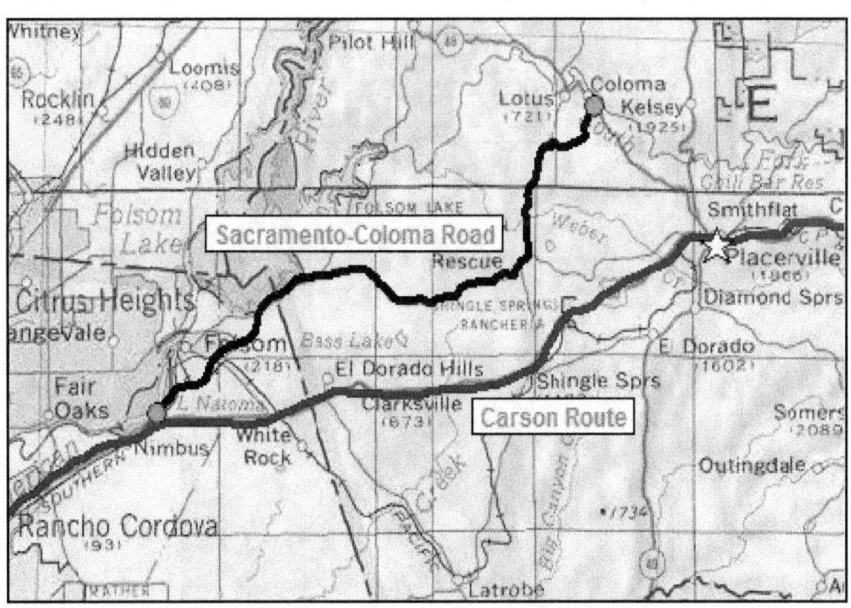

The approximate routes of the Coloma Road (here labeled Sacramento-Coloma Road) and the Sacramento-Placerville Road (here labeled Carson Route and also known as the Carson Immigrant Trail) in El Dorado County that have been superimposed upon a modern map by the National Geographic Society. Note the location of Bass Lake situated between the roads north of the old town of Clarksville.

Morrison House

In the 1850s, what is now Highway 50 was a wagon and stage road known as the Sacramento and Placerville Road. Unlike present-day Highway 50, the old road wound through the hills as it passed through Clarksville and Shingle Springs on its way to Placerville. In the early days of the gold rush, a road house called the American House operated near the present Highway 50 and what is now Old Bass Lake Road, on land that would eventually become the Morrison Ranch.

Alexander Morrison, (October 20, 1816 - August 15, 1893) and his wife Jesse (nee Coatsworth) were born in Scotland,

The headstone of Alexander Morrison marking his grave in the Morrison family cemetery off Bass Lake Road and Highway 50 in El Dorado Hills.

and came to California some time in the 1850s. Morrison purchased the American House land along the Sacramento and Placerville Road and founded the Morrison Ranch. The 1895 Punnett map shows the Morrison Ranch land bisected by the Sacramento-Placerville Road. There on the ranch Andrew and Jessie raised a family of four children, two of whom (possibly twins) died in 1858, approximately a week apart, at the age of five.

The 1880 Federal Census of White Oak Township shows Andrew, his wife Jessie, son Andrew, and daughter Isabella

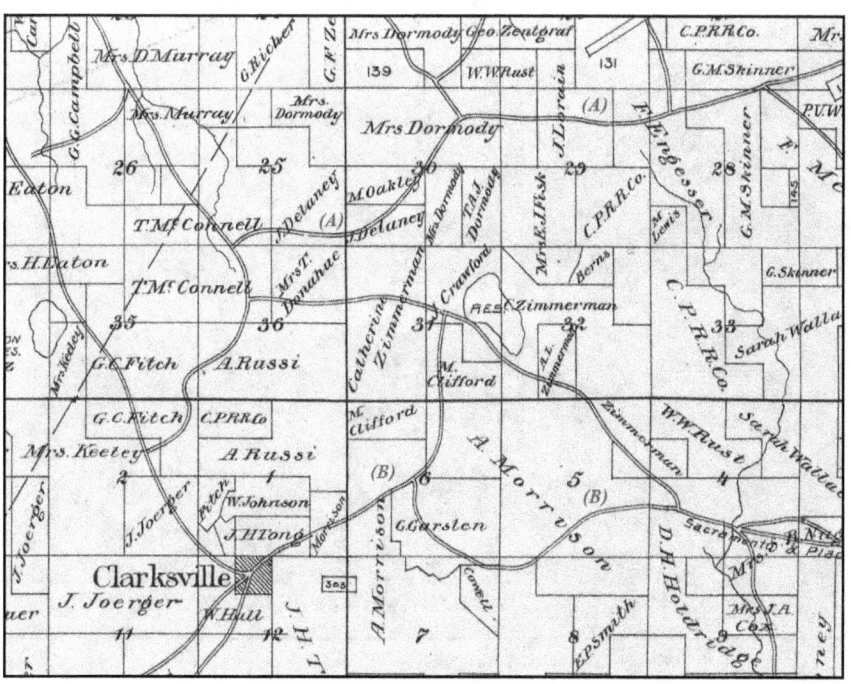

Detail of an El Dorado County map drawn by Punnett in 1895 showing (A) the Old Colma Road traversing the Dormody House property and (B) the Sacramento-Placerville Road traversing the Morrison Ranch property.

living on the Morrison Ranch.

Alexander and Jessie, as well as their last two children, were buried in a small, private cemetery located on their ranch land in the Morrison Ranch Cemetery, also known as the American House Cemetery. Also buried in this cemetery were Jessie's brother Thomas, and Thomas' wife Laurinda, as well as one of Alexander and Jessie's grandchildren. It is believed that Jessie's first two children may also be buried there, but no markers exist to prove that assumption.

A view of the Dormody House, also known as the Green Valley House, as illustrated in Paolo Sioli's 1883 book "Historical Souvenir of El Dorado County, California." The house itself is the same as that sketched by the Rescue Historical Society.

Today, the cemetery, located on a small hill near present-day Highway 50, is still on private property, and is no longer owned by the Morrison family.

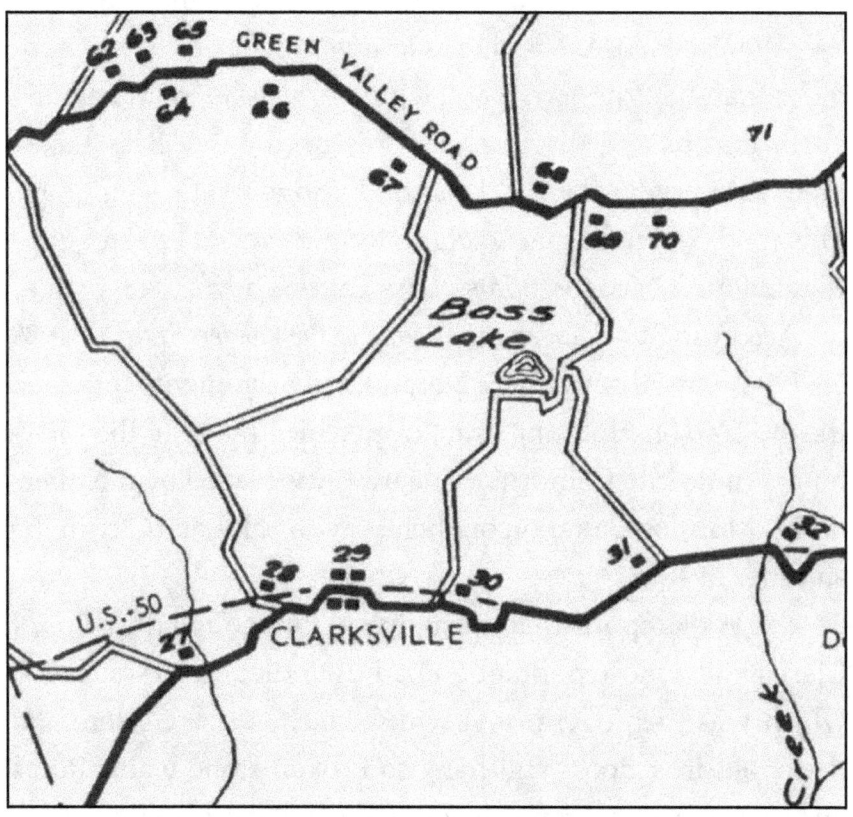

Detail of a map of the early inns of El Dorado County as of July 1859 and drawn by Gene Grueuel in 1954 which appears in the book "I Remember." The map shows Mormon Tavern (28), Mimi Tong's Railroad House in Clarksville (29), Samuel Freeman's Place, Atlantic House (30), and Ohio House (31) along the Placerville Road (now Highway 50). To the south, Green Spring House, later Dormody House (67), Pleasant Grove House (68), and Green Valley House (70) are shown along the Old Coloma Road (now Green Valley Road). Note that the American Reserviour is called Bass Lake on the map.

Atlantic House

Not much is known about Atlantic House. Up the hill towards Placerville and to the east of Margaret Tong's Railroad House at Clarksville, were Samuel Freeman's place and the Atlantic House. Nothing is known about these two stops other than that they were at the junction with a road heading north that passed by American Reservoir (now Bass Lake), then a reservoir for the Diamond Ridge Water Company. This road, which has its northern termination at Green Valley Road, would become today's Bass Lake Road.

According to *California: A Guide to the Golden State,* in 1939 US Highway 50 between El Dorado and Sacramento followed the old Carson Emigrant Trail, over which many of the forty-niners came into California. The trail was blazed by members of the Mormon Battalion on their way to Salt Lake City in the summer of 1848.

The book reports that the highway ran though the ruins of Clarksville, whose population had dwindled to 25 souls, and which was then overgrown with ailanthus trees: "Here and there, old iron doors still hang to broken stone walls. Roofs and windows are gone. Beyond the village signs of abandoned placer diggings are seen in the fields. From here, the broad, treeless hills slope gently westward to the great Sacramento Valley."

Green Springs or Dormody House

Coloma and Sacramento Road was one of the primary routes to the gold fields of El Dorado County. The road, also simply called Coloma Road, ran from Sacramento to Folsom, Mormon Island, Green Valley, Uniontown (now Rescue), then on to Coloma. In 1851 the ferry across the South Fork of the American River was replaced by a truss bridge. Earlier bridges were washed away in floods.

The Green Springs House, an inn and supply stop along the Old Coloma Road, was on the south side of the road near what is now the intersection of Green Valley Road and Deer Creek Road. It was built by Rufus Hitchcock, who had been connected with the Sutter's Fort Hotel. Hitchcock had settled the area in 1848, and he operated the inn until his death from smallpox in 1851.

William Dormody, a native of County Kilkenny, Ireland, and a successful merchant and businessman from the Midwest, arrived in the gold country that same year. Dormody promptly bought Green Springs House and ran it until he died in 1876.

In 1854, Dormody purchased the approximately 1000-acre Green Springs Ranch at auction for $6,400. The ranch was a popular retreat for travelers, especially on hot summer days, and became a favorite location for weddings and other festivities. Along the way, Dormody also opened stores in Georgetown, Coloma, and Kelsey.

In January of 1856, Dormody married the sister of one of his employees, Sarah Francis Norton. Sarah was eighteen when she married the sixty-year-old Dormody.

The couple settled on the Green Springs Ranch and by their tenth anniversary had eight children. In September of 1876 Dormody died from injuries received in an accident when he lost control of his team and his wagon overturned.

Sarah Dormody continued to run the establishment, She

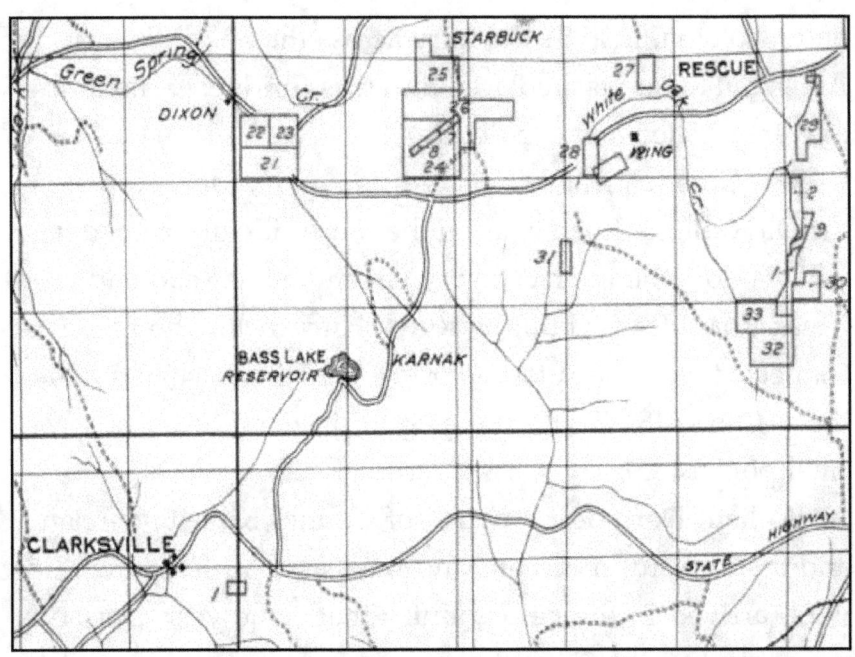

Detail of a map of Western El Dorado County prepared by the California State Bureau of Mines in July of 1938. Note that Bass Lake Road then traversed nearly the route it takes today.

died 26 years later, and was buried beside her late husband in St. John the Baptist Cemetery in Folsom.

In 1956, Howard Greehalgh purchased Green Springs Ranch from the Dormody estate, and in 1976 the property was divided into several parcels. The Green Springs Ranch Rural Development, with 107 ranchettes, sits on one of the largest parcels.

Green Springs Ranch is located at the present intersection of Deer Valley Road and Green Valley Road. It is one of the oldest documented pioneer settlements in El Dorado County, and remnants of the old Coloma Road can still be seen near the entrance to the ranch.

Bass Lake Road

The first mention of the road occured in 1856, when the Board of Supervisors sent Wm. Henry and Joseph Baxter to locate a road that ran from the Morrison House on the Sacramento and Placerville road to intersect the Coloma and Sacramento Road at the Green Valley House in White Oak township, its route passing the reservoir.

It appears that the proposed road passed over the property of one J.G. Gridley because on September 22, 1856, the Board of Supervisors asked Gridley to appear before the Board and show why the road should not be declared a public highway.

Gridley may have successfully resisted the roadway, for on October 6, 1856 the Board of Supervisors dismissed a petition of citizens of White Oak Township for a public highway from the Morrison House on the Placerville and Sacramento Road to the Green Valley House on the Sacramento and Coloma Road.

Later Road Developments

The next mention we find of the road is in 1863. On February 2, 1863: "It is ordered that the road now traveled commencing on the west side of the Ohio House on the Placerville and Sacramento Road and running thence along the west lines of Zimmerman's, Rust's and Willet's Ranches to Evans Store on the Coloma and Folsom Road, be and the same is hereby declared to be a public highway to the width of 60 feet."

Though the portion of this road from the Ohio House to Zimmerman's (Bass Lake) is no longer in existence, apparently few changes have been made in the alignment from the original route.

The road would eventually come to be called Bass Lake Road some time in the 1930s, probably when James Nicol purchased the American Reservoir property from the Diamond Ridge Water Co. Nichol offered bass fishing and other recreation at what he called his Bass Lake Resort located on the eastern shore of the lake. Many old maps however continued to label the reservoir as "American Reservoir."

In the late 1930s the road was aligned further away from the easterly end of Bass Lake due to flooding conditions in the winter.

In the mid-1960s, the southerly end was realigned due to freeway construction. The stub end that was bypassed became Old Bass Lake Road, which one can take from Bass Lake Road to the old Lincoln Highway,

In the 1950s another change in alignment was made at the Mayhew Place to remove the road from in front of the house and to do away with a right angle turn.

In 1978, a fill was constructed across the easterly end of Bass Lake and the road was realigned across this fill.

The source of much of the material in this chapter is the unpublished manuscript, "El Dorado County Road History," by Claibourne W. Trumley, Right of Way Agent, County of El Dorado, May 1980.

~ 16 ~

JAMES NICOL AND
THE BASS LAKE RESORT

Most newcomers to the El Dorado Hills area do not know that Bass Lake got its modern name from the fishing resort that once was operated by James Nicol on the shore of the lake.

James M. Nicol

James Milton Nicol was born in Hillsboro, Trail County, North Dakota on March 14, 1888, the son of John and Elizabeth Nicol (nee Falconer), both of whom were born in the English-speaking part of Canada. James' older siblings, Agnes (1882) and William (1884) ,were also born in Hillsboro.

Sometime around 1893 the Nicol family moved to Koons Street in North Silverton, Marion County, Oregon. His sister Mina (1894) and his brother Clarence (1897) were born there.

In 1910, James and his family were still living in North Silverton. James had finished his elementary education through eighth grade. He had become a steamfitter, working for a plumbing company.

James married Bessie, age 16, in 1914 and they moved to North 26th Street in Billings, Montana. In 1917, at age 29, James was tall and slender, with blue eyes and dark hair, according to his World War I Draft Registration Card. James and Bessie were still married in 1920 but divorced some time thereafter.

James married Eva, age 17, in 1922 and the couple moved to East 19th Street in Oakland, California, where they had two children, Merlyn and Bernal. James and Eva were still living in Oakland in 1930, and James continued to list his occupation as steamfitter. James and Eva probably divorced some time before 1936.

James Nicol Buys American Reservoir

Sometime in the mid-1930s James moved to White Oak Township, which covered the southwestern part of El Dorado County around Clarksville.

On April 27, 1936, we find that James entered into an agreement of sale with the Diamond Ridge Water Company, which may have been a lease or an option on the American Reservoir property, for the purpose of Nicol's starting a fishing resort business at the lake. In any event, Nicol's Bass Lake Resort eventually included picnic facilities, a concession stand that sold refreshments and fishing tackle, and rowboat rentals. James lived in a three-room cabin at Bass Lake Resort.

On June 16, 1938, Nicol bought the American Reservoir and the surrounding property from the Diamond Ridge Water Company for the sum of ten dollars. He also received the ditches and water rights that were west of the Town of Shingle Springs. (El Dorado County Deeds, Book 167: 158)

In August of 1938, Nicol executed a deed of trust with Bank of America. (El Dorado County Deeds, Book 163: 391)

In 1940, James was living at Bass Lake and had a hired man named Joe Knudsen. In 1941, Nicol entered into a grant deed and reconveyance arrangement with E.E. Shields and Herbert Hicks.

James continued to operate the Bass Lake Resort through World War II, listing as his next of kin his sister, Mina, who still lived in Oregon.

Memories of the Bass Lake Resort

Current El Dorado Hills resident Mary Fisher lived in Rio Vista in the late 1940s and had an uncle who ran cattle in the Folsom area. She recalls that she went to Bass Lake to swim one summer in the late 1940s: "What I mostly remember was it being terribly hot. I was sitting in the back seat of the car with my older cousins as we drove over to the lake. Remember, this was in the days when cars were not air-conditioned! When we finally arrived, there was a concession stand, and rowboats you could rent. There were a bunch of dead fish lying around, and they smelled awful."

Madeline Petersen Moseley moved with her family from San Francisco to Clarksville in 1938, where her father purchased the Clarksville general store and gas station on Highway 50, the old Lincoln Highway. She graduated from the Union School in Clarksville in 1941, and from then on she helped her father run the business.

Madeline recalled that the Bass Lake Resort in the 1930s and 1940s was an important local gathering place, since it was the only nearby lake where families could swim, fish and picnic. The American River tended to run dry in the summer, and Folsom Lake was not formed by Folsom Dam until 1952.

Visitors to Bass Lake would stop at the Peterson's gas station on their way to and from the lake, and Madeline said that

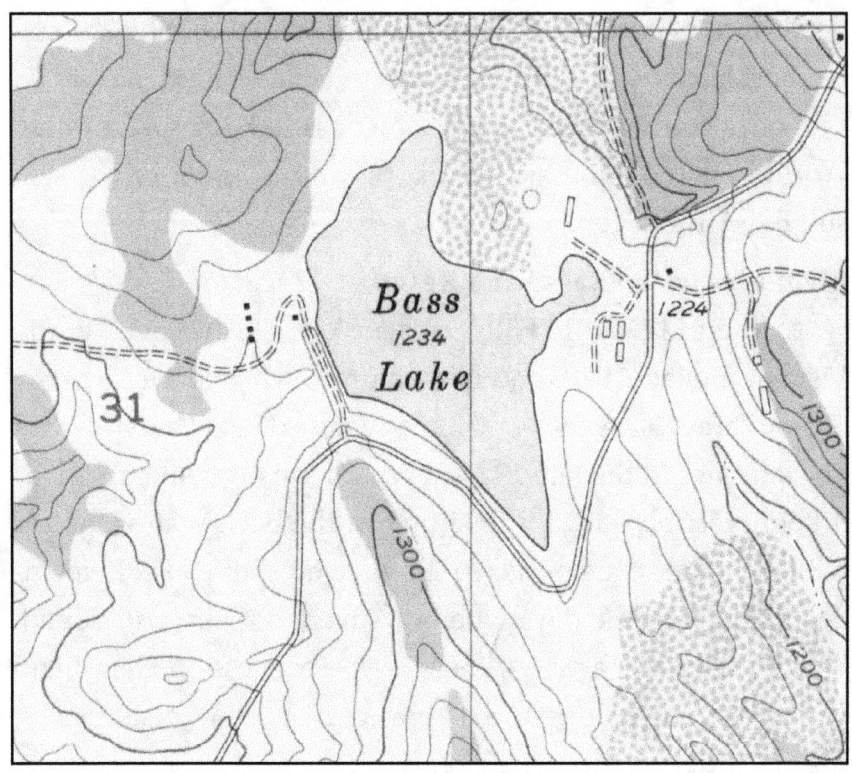

U.S. Geological Survey 1953 Clarksville Quadrangle map, detail of Bass Lake. Map shows what appears to be the Bass Lake Resort on the east side of the lake. Note that Bass Lake Road circles the lake and the current vernal pond is connected to the lake proper.

on weekends folks came from miles around to enjoy themselves at the Bass Lake Resort.

Madeline remembered that there was a store at the lake that sold soft drinks, snacks, and beer, and families would picnic on the ground around the lake shore. Rowboats were also available for fishing or just pleasure rowing on the lake.

Madeline also recalled that Nicol trained hunting dogs at Bass Lake. Nicol would keep the dogs kenneled at the lake,

*James M. Nicol's gravestone at
East Lawn Memorial Park in Sacramento*

and their owners would come up from Sacramento to go possum hunting with the dogs in the area. Occasionally Nicol would bring Madeline a possum to cook for dinner. When folks asked what she was cooking, she said she always told them it was rabbit.

In 1945 Madeline met a serviceman, John Moseley, and they married.

In 1952 Highway 50 was improved and relocated to its present route, bypassing Clarksville, effectively putting the town of Clarksville out of business. Madeline and her husband moved to Folsom, and the store and gas station was closed for ever. "By the time I left Clarksville, I was ready to go," Madeline stated.

James Nicol in the News

The Placerville *Mountain Democrat* of December 28, 1950, reported that Nicol, then 62, was hospitalized in Placerville and being treated for injuries reportedly inflicted upon him by his wife, Doris, 22, who signed a complaint charging him with disturbing the peace. According to the article, the couple got into an altercation on Saturday, December 23, which resulted in both receiving medical treatment. Nicol, described in the article as the owner of Bass Lake and groom of approximately four months, reportedly received 38 stitches in his head to close wounds inflicted by Mrs. Nicol, who wielded a length of oak wood. She claimed that he cursed and struck her and threatened to kill her. The article said that the case was being further investigated. No further trace of Doris Nicol has been found.

Nicol died on October 2, 1952, in the Placerville Sanatorium. The Reverend Henry B. McFadden conducted the final rites in Memory Chapel and interment was in East Lawn Memorial Park in Sacramento.

The assets of Nicol's estate at the Bass Lake Resort were listed as seventeen row boats, a store containing beer, soda, candy, cigars and fishing tackle, and a three-room cabin with household furnishings. Also part of the estate were a 1949 Chevrolet sedan, a house trailer, twenty-eight ewes, two bucks (male sheep), eighteen lambs, and one sack of clipped wool. Nicol was survived by his sister, Mina M. Nicol, and his brother Clarence Nicol, both of Silverton, Oregon.

The American Reservoir property was inherited by James' sister Mina, Mina's daughter Joan Forrest, and James' nephew Bruce Cooper, each of whom received an undivided one-third interest in the property. (El Dorado County Deeds, Book 343:17)

The Walkers' Picnic at Bass Lake

Most of the Serrano development was once the 2,800-acre Walker family ranch. Around 1900, El Dorado Hills was a sleepy ranch community known as Clarksville, settled by immigrants and former miners, most of which raised cattle or sheep. By the late 1800s the more ambitious had grown original homestead claims through marriage, acquisition, or more nefarious means.

Frank Walker was one such man. In a 30-year period surrounding the turn of the last century, Walker expanded his holdings to 2,800 acres, filling his pastures with up to 800 head of cattle bearing his brand, the single letter X. At its peak, the Walker Ranch bordered the Clarksville town site on the south and spanned the area between what is now El Dorado Hills Boulevard and Bass Lake. To the north it bordered the Dixon ranch.

Management of the family ranches passed to Tom Walker, Frank's oldest son, around 1920. At age 25 Tom Walker was a strikingly handsome, well-educated man of means from a prominent family that was well established in both Clarksville and Tahoe City. In 1917 Tom married Eva Belle Miser, a lovely, petite and apparently frail 16-year old from an equally prominent Shingle Springs ranch family.

The couple and their children became active members of the Clarksville community. Patricia "Pat" Walker Johnson, their youngest child, was born in 1929.

Pat Johnson was just 18 months old when her mother succumbed to a bad case of pneumonia two days after Christmas in 1930. Thus began a downward spiral in which Tom Walker would lose his ranch and leave the care of his children to his brother.

Family records indicate that Walker was "apparently devastated" by the loss of his wife, and "seemed unable to manage his young family or the family properties." Tom Walker grew scarce in Clarksville. "Dad just kind of took off, I guess," surmised Johnson. "He was never around."

Tom Walker remains a shadowy figure at this point in the family story, absent to his family, but still involved in the two ranches, apparently making poor decisions. A series of misfortunes led the Walker ranch into decline during the late 1930s.

Walker descendants disagree about where Tom Walker went wrong but generally hold that he was never the same after his first wife died. The Great Depression surely played a part.

After an absence of about two years, Tom Walker unexpectedly showed up at a Walker family picnic at Bass Lake. The remnants of the Walker family reunited with other Clarksville ranch families each year for an annual Mother's Day gathering at the reservoir.

Johnson remembers a well-outfitted couple pulling up in a big car. "Someone said that man was my father," she recalled. "I honestly didn't recognize him."

Tom Walker had a new wife on his arm. Rose Greiner was a hard-working, no-nonsense woman who owned and personally managed three ranches in Woodland. Johnson vividly recalls the look of astonishment on Greiner's face upon meeting her husband's children. Walker had apparently failed to mention his preexisting family to his new wife. Following the awkward introductions, Johnson remembers that Greiner regained her composure and insisted on helping raise the children.

Tom Walker died of cancer in 1949. (Roberts 2012)

Sale of Diamond Ridge Water Company in 1938

As previously mentioned, Nicol bought the American Reservoir and the surrounding property from the Diamond Ridge Water Company on June 16, 1938.

Thirteen days later, June 29, 1938, the Diamond Ridge Water Company, the El Dorado Irrigation District (EID), and the California Door Company (Caldor), entered into an agreement that outlined the conditions of a sale of the Diamond Ridge assets assets to EID.

The agreement specified the manner in which, after the sale, the EID would continue to furnish water to Caldor, and about fifty other existing water customers, for the purposes of various industrial, irrigation, garden, and domestic uses. (El Dorado County Deeds, Book 162: 215)

On October 31, 1938, the Diamond Ridge Water Company sold its assets to EID. However, the American Reservoir property that was sold to James Nicol on June 16, 1938, was specifically excluded from the sale. (El Dorado County Deeds, Book 168: 53)

Evidence of U.S. Geological Survey Maps

The earliest United States Department of the Interior geological map of the Clarksville Quadrangle is dated 1953. That map and all the subsequent USGS maps label what was once called the American Reservoir as Bass Lake.

Subsequent maps show the lake lying along the western side of the north-south Bass Lake Road. The western side of the lake is shown as a straight line, with an unimproved dirt road parallel, which is evidence of the presence of the earthen dam. The map also shows an unimproved road leading to what appear to be several buildings between the road and the

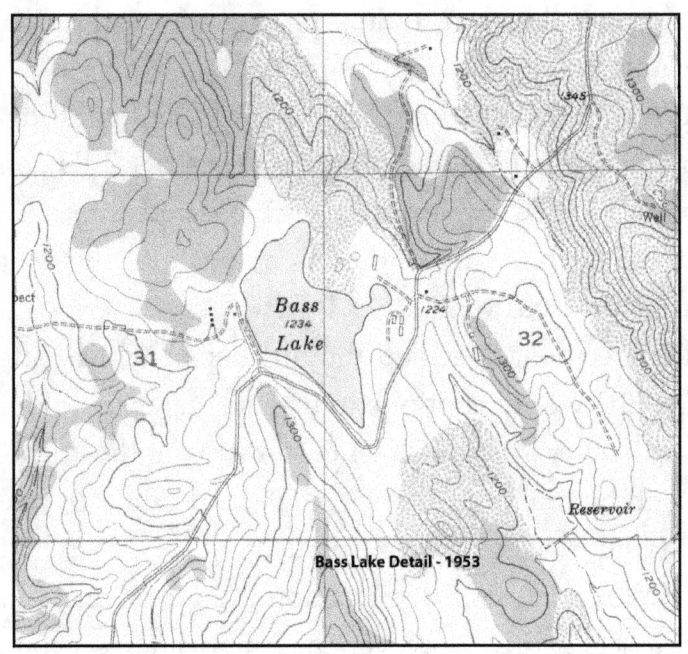

Larger section of the 1953 United States Geological Service map of the Clarksville Quadrant, showing Bass Lake, Bass Lake Road, and what appear to be the roads and buildings of James Nicol's Bass Lake Resort

lake, which were presumably Nicol's Bass Lake Resort.

Sale to Monroe and Valdyne Jannke

On August 4, 1955, the heirs of James Nicol, Mina M. Cooper (nee Nicol), Joan E. Jones (nee Forrest), Bruce Cooper and Carroll J. Cooper (husband and wife), sold the property described as the American Reservoir property to Monroe W. Jannke and Valdyne W. Jannke (husband and wife) of Santa

Cruz, California. (El Dorado County Deeds, Book 343: 17)

Evidence of California Department of Water Resources

New Bass Lake Dam is listed as Dam Number 462 in the California Department of Water Resources Bulletin Number 17 of January 1962, *Dams Within the Jurisdiction of the State of California*. The owner of the dam is listed as Monroe W. Jannke et Ux. According to the publication, the structure of the dam is earthen, its purpose is for storage, and no date of completion is given.

It is not known when the name was changed to New Bass Lake in the California state records. The current designation of the dam in the state records is "New Bass Lake, 53-007."

Sale To El Dorado Irrigation District

On April 21, 1969, Monroe W. Jannke sold the lake known as Bass Lake and the surrounding property to El Dorado Irrigation District. On the deed, the lake was called "American Reservoir (El Dorado County Deeds, Book 927:563). .

The next chapter will discuss the origins of the El Dorado Irrigation District and its activities relating to Bass Lake.

~ 17 ~

EL DORADO IRRIGATION DISTRICT (1960-2018)

In 1960 EID got into the sewage business at the request of Cameron Park leaders, who asked the District to assume operation and maintenance of the community's sewer system.

EID Board members were willing to do so because they viewed recycled water produced at the wastewater treatment plants as a resource. The idea was to use the recycled water, rather than drinking water, for landscape irrigation.

In the years since 1960, the District has constructed, expanded, and renovated many portions of the sewer system to ensure that customers receive reliable service. A separate piped system delivers the recycled water to the yards of approximately 4,000 homes as well as to commercial and public landscapes.

Bass Lake and Recycled Water

Bass Lake played a role in the recycled water operations of the District.

Readers will recall that the earthen dam that created Bass Lake (once called American Reservoir) was built in 1858 by the Eureka Canal Company under the supervision of Lewis B. Harris.

At the time EID bought the reservoir, the dam was 22 feet high, and the top of the dam was at an elevation of 1,240 feet.

The reservoir itself covered 88 acres and was rated at a capacity of 550 acre-feet. (California Department of Water Resources, Bulletin 17, January 1962 and August 1965)

More than a hundred years after the dam was built, the State Department of Dam Safety required EID to lower the capacity to 250 acre-feet, and there it remained as a storage facility until the drought of 1976- 1977.

A new 70-foot-high earth-filled dam, embankment and dikes were constructed in 1977, which raised Bass Lake's capacity to 750 acre-feet. A new treatment facility was also constructed in 1978 that put water into the potable class at the rate of 3-million gallons per day.

El Dorado Hills Specific Plan

The Bass Lake property was included in the El Dorado Hills Specific Plan that was adopted in 1987. The lake and the surrounding parcel were designated as Village R. According to the specific plan document:

> Village R constitutes 157 acres of the El Dorado Irrigation District's Bass Lake water reservoir and water treatment facility. Once used as a recreation area, the lake and surrounding properties are no longer available for public use. The lake is now a potable water storage area for use by EID as a source of gravity-fed domestic water for the El Dorado Hills area. A treatment plant and caretaker's residence are also situated in Village R. In spite of its restricted access, Bass Lake does offer a visual water amenity to the . . . travelers using Bass Lake Road. The lake and surrounding properties also constitute an additional

area of permanent open space which, if feasible, should be returned to public recreational use in the future. No development is proposed for Village R.

Bass Lake Declared Redundant

The role of Bass Lake became smaller and smaller as EID continued to build other modern water storage facilities.

The lake was used for a time to store recycled water produced by EID's Latrobe Road water treatment plant. But it was alleged that the recycled water taken from Bass Lake contained excess minerals that stained concrete sidewalks, so the lake was no longer used for that purpose. It seemed that Bass Lake no longer had a use in EID's water distribution system.

~ 18 ~

RESCUE UNION SCHOOL DISTRICT

In 2009 the EID Board of Directors declared the Bass Lake property surplus to its operations and offered it for sale to any nonprofit organization. The nonprofit stipulation was a result of the requirement that the property be maintained as permanent open space under the terms of the El Dorado Hills Specific Plan.

Bass Lake Purchased by Rescue School District

El Dorado County and Rescue Union School District expressed interest, and the school district made the first offer. In February of 2015 the Rescue Union School District purchased Bass Lake and 58 surrounding acres of land for $300,000 from El Dorado Irrigation District.

Alternate School Site Purchased

The school originally planned to build a 20-acre school site behind Bass Lake Road. A public park was also planned on the property. At the time, district spokespersons cited the price of land as the prime factor in the purchase decision.

In September 2015, seven months after the district closed escrow on the $300,000 Bass Lake property, the school district purchased a second school site property, spending $1.625 million to purchase two parcels of land totaling 21 acres on Sienna Ridge Road near Bass Lake Road to build a future school.

The school district superintendent explained that the new

location at Bass Lake Road and Serrano Parkway was a better fit for the district for several reasons, including better access, and as soon as escrow on the new property was complete, the district would consider what to do with the Bass Lake property.

Bass Lake Offered for Sale

In 2016 the school district decided to divest itself of the Bass Lake property. After a committee representing both the surrounding community and the school district recommended that the property be declared surplus to the district, the school board declared that the property was surplus. The board then offered the property for sale to public agencies, the sale price being limited to the district's original cost of acquiring the property.

Only the El Dorado Hills Community Services District (CSD) wanted to acquire the lake and the property. As part of their offer to buy, the CSD circulated a preliminary plan of how they would use the property that included developing the west side of the lake as a recreation park and leaving the east side of the lake as natural open space with walking trails.

~ 19 ~

EL DORADO HILLS COMMUNITY SERVICES DISTRICT

The Rescue Union School District agreed to sell the Bass Lake property to the El Dorado Hills Community Services District (CSD) in September of 2018.

With the purchase, the CSD took the first step towards their concept of a grand park project at Bass Lake. The CSD envisions building a valuable community resource, which would include athletic fields, an outdoor learning center to study the ecosystem of the lake, and passive and recreational use areas.

Various plans and visions were being discussed at the time this book went to press.

Hopefully the 90-acre lake and its surrounding 100 acres of land will continue to be an asset to the community, in addition to its being an artifact of the California Gold Rush.

APPENDIX

Important Milestones

1849 - Gold is discovered in the millrace at Sutter's Mill, in Coloma, California.

1850 - The readily-mined gold along the creeks and rivers begins to run out.

1851 - Ditch companies formed to bring water to dry diggings.

1852 - James, Furman and Bradley, Berdan ditch companies formed.

1854 - Ditch companies run into finacial difficulties.

1855 - Consolidation of ditch companies begins.

1856 - Eureka Canal Company formed to aquire ditch company assets.

1858 - American Reservoir constructed by Eureka Canal Company as canal terminous.

1875 - Eureka Canal Company sold to Park Canal & Mining Co.

1911 - Park Canal & Mining Co. sells assets, including American Reservoir, to Judge McLaughlin as Trustee for investment group.

1917 - Judge McLaughlin, as Trustee for investment group deeds the Eureka Canal Company, Jones, Bradley and American Reservoir extension main trunk canals or ditches to the Diamond Ridge Water Company.

1925 - El Dorado Irrigation District formed in Placerville and assumed the obligations and assets of the El Dorado Water Company.

1938 - On June 16, Diamond Ridge Water Company sells American Reservoir to Jamed Nicol; Nicol renames American Reservoir "Bass Lake" and starts a fishing resort business there.

1938 - Shortly after the sale of American Reservoir to James Nicol, the bankrupt Diamond Ridge Water Company is acquired by the El Dorado Irrigation District.

1952 - Nicol dies and Bass Lake passes to his heirs.

1955 - Heirs of Nicol sell Bass Lake to the Monroe Jannkes.

1969 - The Jannkes sell Bass Lake to El Dorado Irrigation District.

2015 - El Dorado Irrigation District sells Bass Lake to Rescue Union School District for a school site.

2018 - Rescue Union School District sells Bass Lake to El Dorado Hills Community Services District for recreational and open space use.

BIBLIOGRAPHY

Adams, Frank. *Irrigation Districts in California, Bulletin No. 21,* State of California Public Works, Reports of the Division of Engineering and Irrigation, 1929.

Amador Ledger (Jackson, California). newspaperarchive.com, University of California Riverside, various dates.

Ansted, David T. *The Gold-Seeker's Manual.* London: John Van Voorst, 1849.

Asher and Adams. California and Nevada Map, Holt, 1857.

Ayers, James J. *Gold and Sunshine: Reminiscences of Early California.* Boston: Richard H. Badger, The Gorham Press, 1922.

Blake, William P. *Report Upon the Precious Metals: Being Statistical Notices of the Principal Gold and Silver Producing Regions of the World.* Washington D.C.: Government Printing Office, 1869.

Borthwick, J.D. *The Gold Hunters.* New York: Outing Publishing Company, 1917.

Bostock, John and H.T. Riley. *The Natural History of Pliny, Vol. VI.* London: Henry G. Bohn, 1858.

Bowie, Jesse Augustus. *A Practical Treatise on Hydraulic Mining in California.* New York: D. Van Nostrand, 1885.

Bowman, Amos. *Report on the Properties and Domain of the California Water Company Situated on the Georgetown Divide.* San Francisco: Bancroft & Co. 1874.

Britton & Rey. Map, 1857. Author and Publisher unknown.

Brophy, Alfred L. "Grave Matters: The Ancient Rights of the Graveyard." University of Alabama Public Law Research Paper (August, 2005).

Brown, J. Ross. *Report on the Mineral Resources of the States and Territories West of the Rocky Mountains*. Washington D.C.: Government Printing Office, 1868.

Bielawski, C, J.D. Hoffmann and A. Poett, compilers and publishers, Railroad Map of the Central Part of California and Part of Nevada. Scale 4 miles to the inch. Dated 12 June 1866, rubber stamped 1865.

California Department of Transportation, Cultural Studies Office and JRP Historical Consulting Services. *Water Conveyance Systems in California*. Sacramento, 2000.

California Department of Water Resources. "Bulletin No. 17: Dams Within Jurisdiction of the State of California." Sacramento: January 1962, August 1965, 2014.

California Farmer (San Francisco). newspaperarchive.com, University of California Riverside, various dates.

California Mines and Minerals. Prepared for the meeting of the American Institute of Mining Engineers. San Francisco: California Miners' Association, 1899.

California State Mining Bureau. *Map of El Dorado County, California, Showing Boundaries of the National Forests*. San Francisco, State Mining Bureau, 1900.

Carle, David. *Introduction to Water in California*. Berkeley, Calif.: University of California Press, 2004.

"The Cemetery Lot: Rights and Restrictions," Editors, 109 *University of Pennsylvania Law Review* 378, 1961.

Collins, J.H. *Principles of Metal Mining*. New York: G.P. Putnam's Sons, 1874.

Comstock, J.L. *A History of the Precious Metals from the Earliest Periods to the Present Time*. Hartford: Belknap and Hammersley, 1849.

Condon, W.E. *Sutter, Yuba, Nevada, Placer and El Dorado County*. Map. Sacramento: State Mining Bureau, 1916.

Cox, S. Herbert. *Prospecting for Minerals: A Practical Handbook.* London: Charles Griffin & Company, 1906.

Crane, Walter Richard. *Gold and Silver.* New York: John Wiley & Sons, 1908.

Cronise, Titus Fey. *The Natural Wealth of California.* San Francisco: H.H. Bancroft & Company, 1868.

Cross, Ralph H. *The Early Inns of California, 1844 - 1869,* Cross & Brandt, San Francisco, 1954.

Daily Alta California (San Francisco). newspaperarchive.com, University of California Riverside, various dates.

Del Mar, Alex. *A History of the Precious Metals from the Earliest Times to the Present,* 2d ed. New York: Cambridge Encyclopedia Company, 1902.

Dixon, Christine R. "Deserting God's Acre: The Problem of Abandoned Cemeteries in North Carolina." *Wakeforest Journal of Law & Policy,* 6, 2015, pp.1-18.

Durham, David L. *California's Geographic Names: A Gazetteer of Historic and Modern Names of the State.* Fresno: Quill Driver Books, 1998.

El Dorado County. "El Dorado Hills Specific Plan." Prepared by Wade Associates for the El Dorado County Community Development Department, Placerville, California. Approved by the Board of Supervisors July 18, 1988.

El Dorado Irrigation District. "Serving El Dorado Since 1925: A Brief History of El Dorado Irrigation District." Pamphlet printed by the El Dorado Irrigation District, Placerville, California, 2011.

Emmons, Samuel Franklin and George Ferdinand Becker. *Precious Metal Deposits of the Western United States.* Washington (DC): Government Printing Office, 1885.

Emmons, Samuel Franklin and George Ferdinand Becker. *Statistics and Technology of the Precious Metals*. U.S. Census Bureau, Washington, DC: Government Printing Office, 1885.

Evans, George H. *Practical Notes on Hydraulic Mining.* San Francisco: John Taylor & Co., 1899.

Farish, Thomas E. *The Gold Hunters of California.* Chicago: M.A. Donahue & Co., 1904.

Federal Writers' Project of the Works Progress Administration for the State of California. *California: A Guide to the Golden State* (American Guide Series). Hastings House, New York, 1939.

Final Environmental Impact Report: El Dorado Hills Specific Plan (SCN 86122912). El Dorado County Community Development Department, Jones & Stokes Associates, 1988.

Fisher, Mary. Interview with author, El Dorado Hills, California, 2014.

Frost, John. *History of the State of California, From the Period of Conquest by Spain, to Her Occupation by the United States of America.* Auburn (N.Y.): Derby and Miller, 1851.

Garside, Larry J., Christopher D. Henry, James E. Faulds, and Nicholas H. Hinz. "The upper reaches of the Sierra Nevada auriferous gold channels, California and Nevada." In *Geological Society of Nevada Symposium 2005: Window to the World*, edited by H.N. Rhoden et. al. Reno, Nevada, May 2005.

Gibbes, Charles Drayton. A New Map of the Gold Region in California. New York: Sherman & Smith, 1851.

Gibbs, H.D. and Warren Holt. California and Nevada Map. Philadelphia: S.B. Linton, 1869.

Gibson, J. Watt. *Recollections of a Pioneer.* St. Joseph, Missouri: Nelson-Hanne Printing Co., 1912.

Gold Mines and Mining in California. San Francisco: George Spalding & Co., 1885.

Hanks, Henry G. *Second Report of the State Mineralogist of California from December 1, 1880, to October 1, 1882.* Sacramento: State of California, 1882.

Hanks, Henry G. *Fourth Annual Report of the State Mineralogist for the Year Ending May 15, 1884.* Sacramento: State of California, 1884.

Hittell, John S. *Mining in the Pacific States of North America.* New York: John Wilie, 1862.

Hittel, John S. *Hittel on Gold Mines and Mining of California.* Quebec: G. & G.E. Desbarats, 1864.

Howe & Ferry. Map of Principal Gold Quartz Mines in El Dorado County, California. New York: Howe & Ferry, Printers, *circa* 1850.

Hutchings, James Mason. "A Saw Mill Railroad." *Hutchings California Magazine,* August 1857, 62.

Hume, Fergus. *Madame Midas.* New York: Yurita Press, 2015.

Hoskin, Arthur J. *The Business of Mining.* Philadelphia: J.B. Lippincott Co. 1912.

Howe & Ferry. *Map of the Principal Gold Quartz Mines or Lodes in El Dorado County, California.* New York: Howe & Ferry Press, (undated).

Jackson, Wm. A. *Map of the Mining District of California.* New York: Lambert & Lane's Lith., 1850.

Jefferson, Robert L. *Roughing it in Siberia, with Some Account of the Trans-Siberian Railway, and the Gold-Mining Industry of Asiatic Russia.* London: Sampson Low, Marston & Company, 1897.

Kelly, Dorine. "A Legacy of Bringing water to the People." Placerville, Calif: *Mountain Democrat,* October 4, 1985.

[Kelly, Dorine] "History of El Dorado White Gold." Placerville: El Dorado Irrigation District, 4 October 1985.

Kemp, James Furman. *The Ore Deposits of the United States and Canada,* 4th ed. New York: The Scientific Publishing Company, 1901.

King, C.W. *The Natural History of Precious Stones and of the Precious Metals.* London: Bell & Daldy, 1867.

Krupnik, Vladimir, ed. *History of Gold in Russia.* Moscow, 1994.

Lakes, Arthur. *Prospecting for Gold and Silver.* Scranton: The Colliery Engineer Co., 1895.

Lindgreen Waldemar. *The Tertiary Gravels of the Sierra Nevada of California.* Washington D.C.: United States Geological Survey, 1911.

Lindstrom, Susan, John Wells and Norman Wilson. "Chasing Your Tailings: a Review of Placer Mining Technology." In the Proceedings of the 33rd Annual Meeting of the Society for California Archaeology held in Sacramento April 23-25, 1999, edited by Judyth Reed, 59-83. Fresno: Society for California Archaeology, 2000.

Lock, Alfred G. *Gold: Its Occurrence and Extraction.* London: E. & F.N. Spon, 1882.

Moseley, Madeline Petersen, Interview with author. Folsom, California, 2014.

Map of California 1876. Author and Publisher unknown.

Map of Northern California, 1884. Author and Publisher unknown.

Map of California to Accompany Josiah Royce's *California in American Commonwealths,* 1886, reproduced 2008.

Map of California, 1876a. Author and Publisher unknown.

Map of Western Portion of El Dorado County Showing Mining Claims. 1916 State Mining Bureau, June 1909, San Francisco.

Map of El Dorado County, California, Showing Boundaries of the National Forests. State Mining Bureau, June 1909, San Francisco.

Map of Historic Gold Mines California Department of Conservation, Bureau of Mines, 1997.

Mountain Democrat (Placerville, California). newspaperarchive.com, University of California Riverside, various dates.

Noble, Doug. "Along White Rock Road – Part 2, Clarksville to the Forty Mile House," www.dougstepsout.com, November 22, 2013.

Northern California, Sacramento, El Dorado Counties. Map. California Office of State Engineer, 1884.

Norton, Henry K. *The Story of California From the Earliest Days to the Present,* 7th ed. Chicago: A.C. McClurg & Co., 1924.

Ord, Edward. *Topographical Sketch of the Gold & Quicksilver District of California.* Philadelphia: P.S. Duval's Lith. Steam Press, 1848.

Pacific Rural Press (San Francisco). newspaperarchive.com, University of California Riverside, various dates.

Pany, Beth Knowles. *The People of Clarksville and Its Cemetery.* California Genealogical Society, Oakland, California, 1999.

Patera, Alan H. "El Dorado and Diamond Springs California." Western Places No. 23, Vol. 6, No. 3. Western Places, Lake Grove, Ore. 2001.

Peabody, George W. "Water and Work, Failure and Success." Paper presented at the Gold Rush Conference held at Fort Mason, Golden Gate National Recreation Area, San Francisco, in 1983. From the files of the El Dorado County Historical Museum, Placerville.

Peabody, George W. "J.J. Crawford and the Crawford Ditch: The Prosperity of Diamond Ridge." Article written for the *Gold Oak Gazette,* Gold Oak School, Placerville, California, circa 1985. From the files of the El Dorado County Historical Museum, Placerville.

Phillips, J. Arthur. *The Mining and Metallurgy of Gold and Silver.* London, E. and F.N. Spon, 1867.

Punnet Bros. of San Francisco, Map of the County of El Dorado, California, compiled for Shelly Inch, Stationer, Placerville, 1895.

Rand McNally and Company. *Commercial Atlas of America,* Map of California (Northern Section). Chicago: Rand McNally and Company, 1924.

Register of Mines and Minerals, El Dorado County, California. San Francisco: State Mining Bureau, April,1902

Report of the Auditor General on the Finances of the Commonwealth of Pennsylvania for the Year Ending November 30 1886. Harrisburg: Edwin K. Meyers, State Printer, 1886.

Report of the Railroad Commission of the State of California, from July 1, 1920, to June 30, 1921. Sacramento: California State Printing Office, 1922.

Roberts, Mike. "The Rise and Fall of The Walker Ranch," *El Dorado Hills Village Life,* May 6, 2012.

Sacramento Daily Union. newspaperarchive.com, University of California Riverside, various dates.

San Francisco Daily Call. newspaperarchive.com, University of California Riverside, various dates.

Seidemann, Ryan M. *How do We Deal With All the Bodies? A Review of Recent Cemetery and Human Remains Legal Issues.* Berkeley: Bepress, 2013.

Seideman, Ryan M. and Rachel L. Moss. *Places Worth Protecting: A Legal Guide to the Protection of Historic Cemeteries in Louisiana and Recommendations for Additional Protection.* Berkeley: Bepress 2009

Shaffer, C. Allen. "Standing of the Dead: Solving the Problem of Abandoned Graveyards." *Capital University Law Review* 32 (2004) 479-498.

Shell Oil Company. Shell Highway Map of California (northern Portion) 6-DD-1956-4. Chicago: Shell Oil Company, 1956.

Shinn, Charles H. *Mining Camps: A Study in American Frontier Government.* New York: Charles Scribner's Sons, 1885.

Sioli, Paolo. *Historical Souvenir of El Dorado County California.* Oakland: Paolo Sioli, 1883.

Starns, Jean E. "Historic Mining Ditches of El Dorado County, California." In the Proceedings of the 33rd Annual Meeting of the Society for California Archaeology held in Sacramento April 23-25, 1999, ed. Judyth Reed, 54-58. Fresno: Society for California Archaeology, 2000.

Starns, Jean E. *Wealth From Gold Rush Waters.* Georgetown (Calif.): Jean E. Starns, 2004.

Strangstad, Lynette. *A Graveyard Preservation Primer.* 2d ed. Lanham, Md: Alta Mira Press, 2013.

Story of Water: From Miner's Ditch to Sly Park Dam, in Financial Report of El Dorado Irrigation District 1951-1952. Placerville, Calif: El Dorado Irrigation District, 1953.

Swinehart, D. Bruce, Jr., American River College. Comment Letter to El Dorado County Department of Transportation, December 27, 1991, on the Environmental Impact Report for Silver Springs Parkway, SCR No. 1991122014.

"To find gold . . . you need water." EID Anniversary Edition 1925-2000 (Supplement). Placerville: *Mountain Democrat* September, 2000.

Trumley, Claibourne W. "El Dorado County Road History." Unpublished manuscript. Placervile: County of El Dorado, 1980.

United States Geological Survey. Map, *Clarksville Quadrangle, California*. Washington, D.C.: Geological Survey, 1955, 1960, 1972, 1976, 1980a, 2012.

Upton, Charles Elmer. *Pioneers of El Dorado*. Placerville: Charles Elmer Upton, 1906.

Water Resources Development and Management Plan. El Dorado County Water Agency. Shingle Springs, Calif.: 2007.

West, J.M. "Bureau of Mines Information, Circular 8517." Washington (DC): U.S. Bureau of Mines, 1971.

Wilson, Eugene B. *Hydraulic and Placer Mining*, 3d ed. New York: John Wiley & Sons, Inc., 1918.

Woolbridge, Jesse Walton. *History of the Sacramento Valley* vol 2. Chicago: The Pioneer Historical Pub. Co., 1931.

Wyld, James. *Map of the Gold Regions of California*. London: James Wyld, 1849.

Yohalem, Betty. *"I Remember . . .": Stories and Pictures of El Dorado County Pioneer Families* (1st ed.). Placerville: El Dorado Chamber of Commerce, 1977.

About the Author

John E. Thomson, a native of California, is a founding member of the Clarksville Region Historical Society, and has been a member of its Board of Directors for a number of years.

During his professional career John held research or management positions at GATX Leasing Corporation, San Francisco; Bechtel Corporation, San Francisco; and Syntex Corporation, Palo Alto. He consulted for Cisco Sytems, Raychem Corporation, and Broken Hill Proprietary Company Limited (BHP).

In academia, John held various positions at Phillips Junior College, Heald College, National Hispanic University, John Moores University (Liverpool), and the Hoover Institution at Stanford University. He earned an Associate of Arts (A.A.) from El Camino College, Torrance, California; a Bachelor of Science in Administrative Science (B.S.A.S.) from Pepperdine University, Malibu; a Master of Business Administration (M.B.A.) from Golden Gate University, San Francisco; and a Doctor of Philosophy (Ph.D.) in International Business from International University of America, London.

John is the author of numerous papers and monographs, including "Entertainment Services on the Information Highway: The Case for Video-on-Demand" (1996), and co-authored the book *Life After Layoff* (1999) with his wife Frances Thomson.

John and Fran live with their cat, Miss Puss, in El Dorado Hills, California.

www.ingramcontent.com/pod-product-compliance
Lightning Source LLC
Chambersburg PA
CBHW050554300426
44112CB00013B/1915